Noise Pollution: Causes, Effects and Management

Noise Pollution: Causes, Effects and Management

Julia Page

SYRAWOOD
PUBLISHING HOUSE
New York

Published by Syrawood Publishing House,
750 Third Avenue, 9th Floor,
New York, NY 10017, USA
www.syrawoodpublishinghouse.com

Noise Pollution: Causes, Effects and Management
Julia Page

International Standard Book Number: 978-1-64740-041-5 (Hardback)

Cataloging-in-Publication Data

Noise pollution : causes, effects and management / Julia Page.
p. cm.
Includes bibliographical references and index.
ISBN 978-1-64740-041-5
1. Noise pollution. 2. Pollution. 3. Pollution prevention.
4. Environmental protection I. Page, Julia.
TD892 .N65 2020
620.23--dc23

Printed in China.

Table of Contents

Preface

This book aims to help a broader range of students by exploring a wide variety of significant topics related to this discipline. It will help students in achieving a higher level of understanding of the subject and excel in their respective fields. This book would not have been possible without the unwavered support of my senior professors who took out the time to provide me feedback and help me with the process. I would also like to thank my family for their patience and support.

The propagation of noise that can have a harmful impact on the activities of human and animal life is known as noise pollution. Major sources of noise pollution include machines, transport and propagation systems. It can also be caused by loud music, construction and electrical generators. High levels of noise pollution can give rise to cardiovascular effects in humans and increase the incidences of coronary artery disease. Noise can also have a detrimental effect on animals. It can upset the delicate balance in predator or prey detection and avoidance, as well as interfere in the use of sound for communication. Different approaches, evaluations and methodologies and advanced studies on noise pollution have been included in this book. It covers in detail some existent theories and innovative concepts revolving around this problem. The book is appropriate for those seeking detailed information in this area.

A brief overview of the book contents is provided below:

Chapter – Introduction

The unwanted sound that is judged to be loud or disruptive to hearing is referred to as noise. The propagation of noise that has a harmful impact on the activity of human or animal life is known as noise pollution. This is an introductory chapter which will introduce briefly all the significant aspects of noise and noise pollution.

Chapter – Classification of Noise

Noise can be classified into various types. Some of these are shot noise, atmospheric noise, Brownian noise, burst noise, flicker noise, Gaussian noise, Johnson–Nyquist noise and grey noise. The topics elaborated in this chapter will help in gaining a better perspective about these types of noise.

Chapter – Types of Noise pollution

On the basis of the noise type and its origin, noise pollution can be classified into various types. Some of them are roadway noise, occupational noise, aircraft noise pollution, loud music, environmental noise, train noise, etc. All these types of noise pollution have been carefully analyzed in this chapter.

Chapter – Effects of Noise Pollution

Noise has many harmful and disturbing effects on animals and humans. The major effects of noise pollution in humans include acoustic trauma and noise-induced hearing loss. This chapter has been carefully written to provide an easy understanding of the various adverse effects of noise on humans as well as wildlife.

Chapter – Noise Pollution: Reduction and Control

Noise pollution can be controlled using various techniques and systems such as soundproofing, radio noise reduction system, acoustiblok, noise barrier, muzzle shroud, hearing protection device, hush kit, acoustic quieting, etc. The chapter closely examines these methods of controlling and reducing noise.

Chapter – Sound Level Measurement

Various instruments are used to for the measurement of sound level. The most common of them are microphones, sound level meters, frequency analyzers, noise dosimeters, personal sound level meters, calibrators, etc. All these diverse methods and instruments of sound level measurement have been carefully discussed in this chapter.

Julia Page

1

Introduction

The unwanted sound that is judged to be loud or disruptive to hearing is referred to as noise. The propagation of noise that has a harmful impact on the activity of human or animal life is known as noise pollution. This is an introductory chapter which will introduce briefly all the significant aspects of noise and noise pollution.

NOISE

Noise can be defined as "unwanted sound", and an audible acoustic energy that adversely affects the physiological and/or psychological well-being of people, or which disturbs or impairs the convenience or peace of any person. We can generalise by saying that sound becomes unwanted when it:

- Hinders speech communication.
- Impedes the thinking process.
- Interferes with concentration.
- Obstructs activities (work or leisure).
- Presents a health risk due to hearing damage.

It is important to remember that whether a given sound is "noise" depends on the listener or hearer. The driver playing loud rock music on their car radio hears no noise, but the person in the traffic behind them hears nothing but noise.

NOISE POLLUTION

Noise pollution takes place when there is either an excessive amount of noise or an unpleasant sound that causes a temporary disruption in the natural balance. This definition is usually applicable to sounds or noises that are unnatural in either their volume or their production. Our environment is such that it has become difficult to escape the noise. Even electrical appliances at home have a constant hum or beeping sound. By and large, lack of urban planning increases the exposure to unwanted sounds. This is why understanding noise pollution is necessary to curb it in time.

Various Causes of Noise Pollution

Industrialization

Most of the industries use big machines which are capable of producing a large amount of noise. Apart from that, various equipment like compressors, generators, exhaust fans, grinding mills also participates in producing big noise. Therefore, you must have seen workers in these factories and industries wearing earplugs to minimize the effect of noise.

Poor Urban Planning

In most of the developing countries, poor urban planning also plays a vital role. Congested houses, large families sharing small space, fight over parking, frequent fights over basic amenities leads to noise pollution which may disrupt the environment of society.

Social Events

Noise is at its peak in most of the social events. Whether it is marriage, parties, pub, disc or place of worship, people normally flout rules set by the local administration and create nuisance in the area. People play songs on full volume and dance till midnight which makes the condition of people living nearby pretty worse. In markets, you can see people selling clothes via making a loud noise to attract the attention of people.

Transportation

A large number of vehicles on roads, airplanes flying over houses, underground trains produce heavy noise and people get it difficult to get accustomed to that. The high noise leads to a situation wherein a normal person loses the ability to hear properly.

Construction Activities

Under construction activities like mining, construction of bridges, dams, buildings, stations, roads, flyovers takes place in almost every part of the world. These construction activities take place every day as we need more buildings, bridges to accommodate more people and to reduce traffic congestion. The down point is that these construction equipment are too noisy.

Household Chores

We people are surrounded by gadgets and use them extensively in our daily life. Gadgets like TV, mobile, mixer grinder, pressure cooker, vacuum cleaners, washing machine and dryer, cooler, air conditioners are minor contributors to the amount of noise that is produced but it affects the quality of life of your neighborhood in a bad way.

While this form of pollution may seem harmless, it, in fact, has far-reaching consequences. The adverse effects on the health of the environment are quite severe. Not only is the local wildlife affected by pollution but humans also face a number of problems due to it.

2

Classification of Noise

Noise can be classified into various types. Some of these are shot noise, atmospheric noise, Brownian noise, burst noise, flicker noise, Gaussian noise, Johnson–Nyquist noise and grey noise. The topics elaborated in this chapter will help in gaining a better perspective about these types of noise.

SHOT NOISE

Shot noise is a form of noise that arises because of the discrete nature of the charges carried by charge carriers, electrons or holes.

When looking at what is shot noise, it can be seen that it is particularly obvious when current levels are low. This is because the statistical nature of the current flow together with the discrete charge levels is more apparent.

Shot noise is particularly noticeable in semiconductor devices, such as tunnel junctions, Schottky barrier diodes and PN junctions.

The first understanding of shot noise was developed by Walter Schottky, the German physicist who played a major role in the development of the theory of electron and ion emission.

While working on vacuum tubes/thermionic valves, Schottky saw that even when all external sources of noise had been eliminated two types of noise remained. One he determined was a result of the temperature and this is now referred to as thermal noise. The other was shot noise.

Shot noise arises because current consists of a vast number of discrete charges, and is not a totally analogue phenomenon.

The continuous flow of these discrete pulses gives rise to almost white noise. There is a cut-off frequency which is governed by the time it takes for the electron or other charge carrier to travel through the conductor.

Unlike thermal noise, this noise is dependent upon the current flowing and has no relationship to the temperature at which the system is operating.

Shot noise is more apparent in devices such as PN tunnelling junctions. The electrons are transmitted randomly and independently of each other.

For metallic resistors, shot noise is virtually non-existent because the inelastic electron-phonon scattering smoothes the current fluctuations that result from the discrete nature of the electrons, leaving only thermal noise.

However recent developments have recognised that there is still some very small level of shot noise.

When developing low noise amplifiers for RF or audio applications, it helps to be aware of the presence of shot noise so that its effects can be minimised as far as possible in the design.

ATMOSPHERIC NOISE

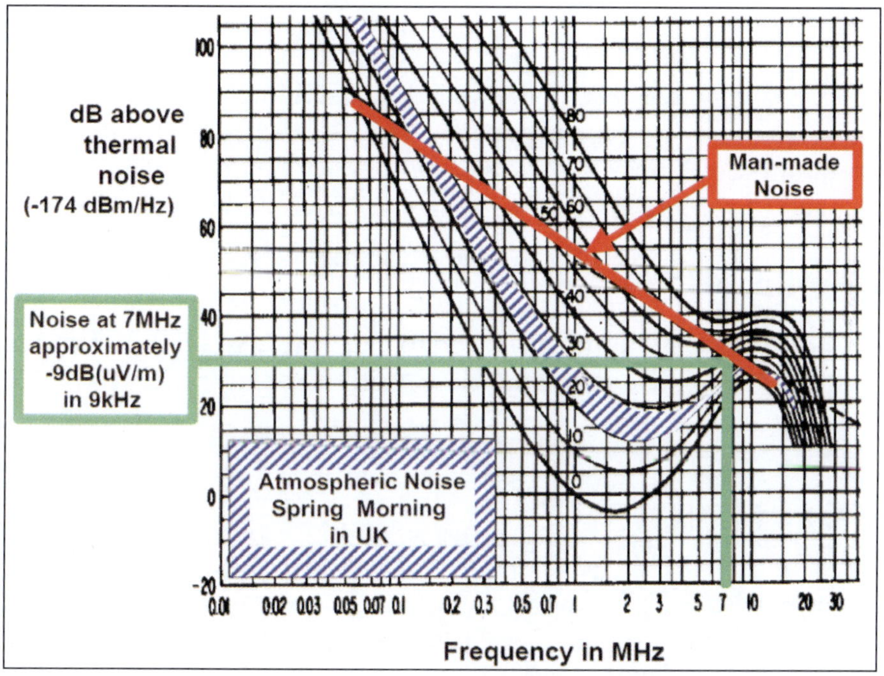

In above figure, CCIR 322 atmospheric noise relationship. The standard has tables and maps that determine the noise figure at 1 MHz according to the season and the time of day. This graph converts that noise figure to other frequencies. Notice that the plotted lines are spaced in 10 dB increments at 1 MHz.

Atmospheric noise is radio noise caused by natural atmospheric processes, primarily lightning discharges in thunderstorms. On a worldwide scale, there are about 40 lightning flashes per second – ≈3.5 million lightning discharges per day.

Lightning

Atmospheric noise is radio noise caused by natural atmospheric processes, primarily lightning discharges in thunderstorms. It is mainly caused by cloud-to-ground flashes as the current is much stronger than that of cloud-to-cloud flashes. On a worldwide scale, 3.5 million lightning flashes occur daily. This is about 40 lightning flashes per second.

The sum of all these lightning flashes results in atmospheric noise. It can be observed, with a radio receiver, in the form of a combination of white noise (coming from distant thunderstorms) and impulse noise (coming from a near thunderstorm). The power-sum varies with seasons and nearness of thunderstorm centers.

Although lightning has a broad-spectrum emission, its noise power increases with decreasing frequency. Therefore, at very low frequency and low frequency, atmospheric noise often dominates, while at high frequency, man-made noise dominates in urban areas.

Survey

From 1960s to 1980s, a worldwide effort was made to measure the atmospheric noise and variations. Results have been documented in CCIR Report 322. CCIR 322 provided seasonal world maps showing the expected values of the atmospheric noise figure F_a at 1 MHz during four hour blocks of the day. Another set of charts relates the F_a at 1 MHz to other frequencies. CCIR Report 322 has been superseded by ITU P.372 publication.

Random Number Generation

Atmospheric noise and variation is also used to generate high quality random numbers. Random numbers have interesting applications in the security domain.

BROWNIAN NOISE

In science, Brownian noise, also known as Brown noise or red noise, is the kind of signal noise produced by Brownian motion, hence its alternative name of random walk noise. The term "Brown noise" does not come from the color, but after Robert Brown, the discoverer of Brownian motion. The term "red noise" comes from the "white noise"/"white light" analogy; red noise is strong in longer wavelengths, similar to the red end of the visible spectrum.

The graphic representation of the sound signal mimics a Brownian pattern. Its spectral density is inversely proportional to f^2, meaning it has more energy at lower frequencies, even more so than pink noise. It decreases in power by 6 dB per octave (20 dB per decade) and, when heard, has a "damped" or "soft" quality compared to white and pink noise. The sound is a low roar resembling a waterfall or heavy rainfall.

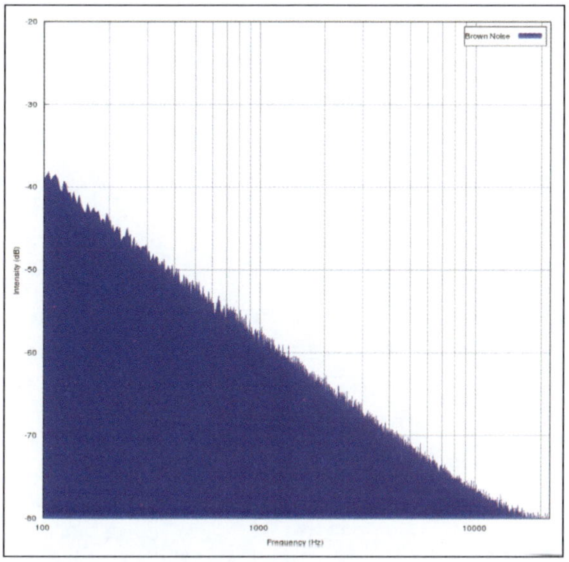

Spectrum of Brownian noise, with a slope of −20 dB per decade.

Strictly, Brownian motion has a Gaussian probability distribution, but "red noise" could apply to any signal with the $1/f^2$ frequency spectrum.

Power Spectrum

A Brownian motion, also called a Wiener process, is obtained as the integral of a white noise signal:

$$W(t) = \int_0^t \frac{dW(\tau)}{d\tau} d\tau,$$

meaning that Brownian motion is the integral of the white noise $dW(t)$, whose power spectral density is flat:

$$S_0 = \left| \mathcal{F}\left[\frac{dW(t)}{dt} \right](\omega) \right|^2 = \text{const.}$$

Here $\mathcal{F}$ denotes the Fourier transform, and S_0 is a constant. An important property of this transform is that the derivative of any distribution transforms as,

$$\mathcal{F}\left[\frac{dW(t)}{dt} \right](\omega) = i\omega \mathcal{F}[W(t)](\omega),$$

from which we can conclude that the power spectrum of Brownian noise is,

$$S(\omega) = |\mathcal{F}[W(t)](\omega)|^2 = \frac{S_0}{\omega^2}.$$

An individual Brownian motion trajectory presents a spectrum $S(\omega) = S_0 / \omega^2$, where the amplitude S_0 is a random variable, even in the limit of an infinitely long trajectory.

Production

Brown noise can be produced by integrating white noise. That is, whereas (digital) white noise can be produced by randomly choosing each sample independently, Brown noise can be produced by adding a random offset to each sample to obtain the next one. A leaky integrator might be used in audio applications to ensure the signal does not "wander off". While the first sample is random across the entire range that the sound sample can take on, the remaining offsets from there on are a tenth or there abouts, leaving room for the signal to bounce around.

BURST NOISE

Burst noise, or as it is sometimes called, popcorn noise, or random telegraph signal, RTS, consists of sudden step-like transitions between two or more levels.

The burst noise steps may be as high as several hundred microvolts, at random and unpredictable times. Each shift in offset voltage or current can last for several milliseconds, and the intervals between pulses tend to be in the low audio range - typically less than about 100 Hz.

Burst noise, or popcorn noise was an issue when the first operational amplifiers were introduced. It made a noise like cooking popcorn if sent to a loudspeaker - hence the name.

The most common cause for this noise in ICs is believed to be the random trapping and release of charge carriers at thin film interfaces. Also defect sites in bulk semiconductor crystal can give rise to burst noise. In some cases the effect can have a greater effect than others. It is particularly severe when it affects an MOS gate or in a bipolar base region. In these cases the output effect can be significant.

These defects can be caused by manufacturing process issues: e.g. heavy ion implantation, or unintentional side-effects such as surface contamination. As the effect is a result of a manufacturing issue, offending ICs can be detected and removed during manufacture.

Owing to the high standards used in semiconductor manufacturing these days, burst noise is not at all common. However when it does occur, it can be difficult to trace and track down, and as a result it is worth understanding that it exists and the mechanism behind it.

FLICKER NOISE

Flicker noise is a type of electronic noise with a 1/f power spectral density. It is therefore often referred to as 1/f noise or pink noise, though these terms have wider definitions. It occurs in almost all electronic devices and can show up with a variety of other effects, such as impurities in a conductive channel, generation and recombination noise in a transistor due to base current, and so on.

Properties

$1/f$ noise in current or voltage is usually related to a direct current, as resistance fluctuations are transformed to voltage or current fluctuations by Ohm's law. There is also a $1/f$ component in resistors with no direct current through them, likely due to temperature fluctuations modulating the resistance. This effect is not present in manganin, as it has negligible temperature coefficient of resistance.

In electronic devices, it shows up as a low-frequency phenomenon, as the higher frequencies are overshadowed by white noise from other sources. In oscillators, however, the low-frequency noise can be mixed up to frequencies close to the carrier, which results in oscillator phase noise.

Flicker noise is often characterized by the corner frequency f_c between the region dominated by the low-frequency flicker noise and the higher-frequency "flat-band" noise. MOSFETs have a higher f_c (can be in the GHz range) than JFETs or bipolar transistors, which is usually below 2 kHz for the latter.

It typically has a Gaussian distribution and is time-reversible. It is generated by a linear mechanism in resistors and FETs, but a non-linear mechanism in BJTs and diodes.

The flicker-noise voltage power in MOSFET is often modeled as $\frac{K}{C_{ox} \cdot WLf}$, where K is the process-dependent constant, C_{ox} is the oxide capacitance in MOSFET devices, W and L are channel width and length respectively. This is an empirical model and generally thought to be an oversimplification.

Flicker noise is found in carbon-composition resistors and in thick-film resistors, where it is referred to as *excess noise*, since it increases the overall noise level above the thermal noise level, which is present in all resistors. In contrast, wire-wound resistors have the least amount of flicker noise. Since flicker noise is related to the level of DC, if the current is kept low, thermal noise will be the predominant effect in the resistor, and the type of resistor used may not affect noise levels, depending on the frequency window.

Measurement

The measurement of $1/f$ noise spectrum in voltage or current is done in the same way as the measurement of other types of noises. Sampling spectrum analyzers take a finite-time sample from the noise and calculate the Fourier transform by FFT algorithm. Then, after calculating the squared absolute value of the Fourier spectrum, they calculate its average value by repeating this sampling process by a sufficiently large number of times. The resulting pattern is proportional to the power-density spectrum of the measured noise. It is then normalized by the duration of the finite-time sample and also by a numerical constant in the order of 1 to get its exact value. This procedure gives correct spectral data only deeply within the frequency window determined by the reciprocal of the duration of the finite-time sample (low-frequency end) and the digital sampling rate of the noise (high-frequency end). Thus the upper and the lower half decades of the obtained power density spectrum are usually discarded from the spectrum. Conventional spectrum analyzers that sweep a narrow filtered band over the signal have good signal-to-noise ratio (SNR), since they are narrow-band instruments. Unfortunately, these instruments do not operate at frequencies low enough to fully measure flicker noise. Sampling instruments are broadband, and hence high noise. They reduce the noise by taking multiple sample traces and averaging them. Conventional spectrum analyzers still have better SNR due to their narrow-band acquisition.

Removal in Instrumentation and Measurements

For DC measurements $1/f$ noise can be particularly troublesome, as it is very significant at low frequencies, tending to infinity with integration/averaging at DC. At very low frequencies, you can think of the noise as becoming drift, although the mechanisms causing drift are usually distinct from flicker noise.

One powerful technique involves moving the signal of interest to a higher frequency and using a phase-sensitive detector to measure it. For example, the signal of interest can be chopped with a frequency. Now the signal chain carries an AC, not DC, signal. AC-coupled stages filters out the DC component; this also attenuates the flicker noise. A synchronous detector that samples the peaks of the AC signal, which are equivalent to the original DC value. In other words, first the low-frequency signal is shifted to high frequency by multiplying it with high-frequency carrier, and it is given to the device affected by the flicker noise. The output of the device is again multiplied with the same carrier, so the previous information signal comes back to baseband, and flicker noise will be shifted to higher frequency, which can easily be filtered out.

GAUSSIAN NOISE

Gaussian noise, named after Carl Friedrich Gauss, is statistical noise having a probability density function (PDF) equal to that of the normal distribution, which is also known as the Gaussian distribution. In other words, the values that the noise can take on are Gaussian-distributed.

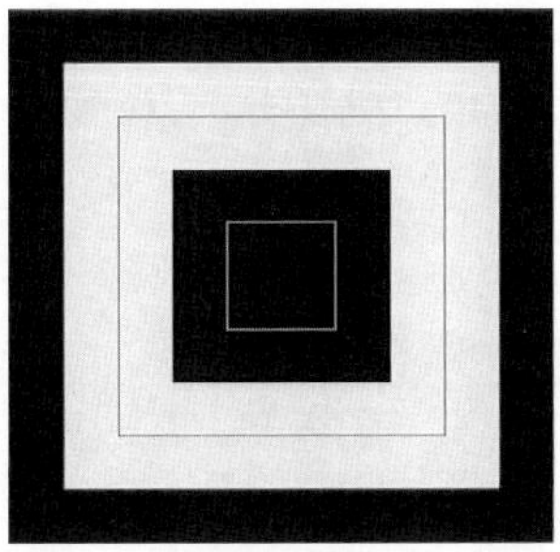

Without noise.

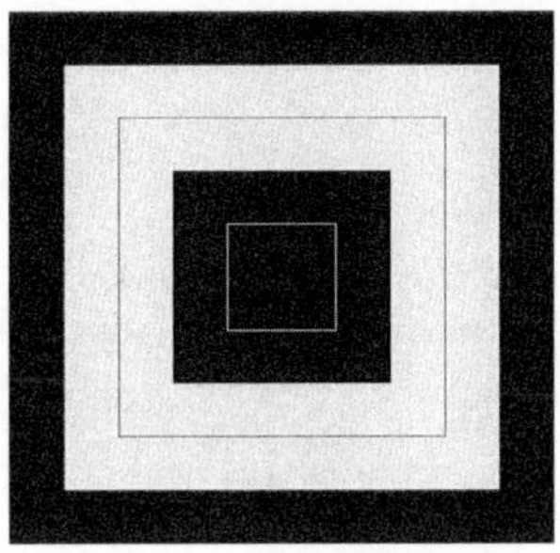

With Gaussian noise.

The probability density function p of a Gaussian random variable z is given by:

$$p_G(z) = \frac{1}{\sigma\sqrt{2\pi}} e^{-\frac{(z-\mu)^2}{2\sigma^2}},$$

where, z represents the grey level, μ the mean value and σ the standard deviation.

A special case is *White Gaussian noise,* in which the values at any pair of times are identically distributed and statistically independent (and hence uncorrelated). In communication channel testing and modelling, Gaussian noise is used as additive white noise to generate additive white Gaussian noise.

In telecommunications and computer networking, communication channels can be affected by wideband Gaussian noise coming from many natural sources, such as the thermal vibrations of atoms in conductors (referred to as thermal noise or Johnson–Nyquist noise), shot noise, black-body radiation from the earth and other warm objects, and from celestial sources such as the Sun.

Gaussian Noise in Digital Images

Principal sources of Gaussian noise in digital images arise during acquisition e.g. sensor noise caused by poor illumination and/or high temperature, and/or transmission e.g. electronic circuit noise. In digital image processing Gaussian noise can be reduced using a spatial filter, though when smoothing an image, an undesirable outcome may result in the blurring of fine-scaled image edges and details because they also correspond to blocked high frequencies. Conventional spatial filtering techniques for noise removal include: mean (convolution) filtering, median filtering and Gaussian smoothing.

JOHNSON–NYQUIST NOISE

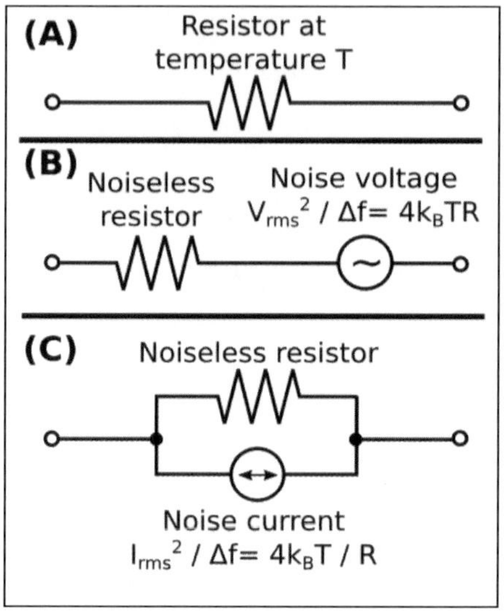

These three circuits are all equivalent: (A) A resistor at nonzero temperature, which has Johnson noise; (B) A noiseless resistor in series with a noise-creating voltage source (i.e. the Thévenin equivalent circuit); (C) A noiseless resistance in parallel with a noise-creating current source (i.e. the Norton equivalent circuit).

Johnson–Nyquist noise (thermal noise, Johnson noise, or Nyquist noise) is the electronic noise generated by the thermal agitation of the charge carriers (usually the electrons) inside an electrical conductor at equilibrium, which happens regardless of any applied voltage. Thermal noise is present in all electrical circuits, and in sensitive electronic equipment such as radio receivers can drown out weak signals, and can be the limiting factor on sensitivity of an electrical measuring instrument. Thermal noise increases with temperature. Some sensitive electronic equipment such as radio telescope receivers are cooled to cryogenic temperatures to reduce thermal noise in their circuits. The generic, statistical physical derivation of this noise is called the fluctuation-dissipation theorem, where generalized impedance or generalized susceptibility is used to characterize the medium.

Thermal noise in an ideal resistor is approximately white, meaning that the power spectral density is nearly constant throughout the frequency spectrum. When limited to a finite bandwidth, thermal noise has a nearly Gaussian amplitude distribution.

Noise Voltage and Power

Thermal noise is distinct from shot noise, which consists of additional current fluctuations that occur when a voltage is applied and a macroscopic current starts to flow. For the general case, the above definition applies to charge carriers in any type of conducting medium (e.g. ions in an electrolyte), not just resistors. It can be modeled by a voltage source representing the noise of the non-ideal resistor in series with an ideal noise free resistor.

The one-sided power spectral density, or voltage variance (mean square) per hertz of bandwidth, is given by,

$$\overline{v_n^2} = 4k_{\mathrm{B}}TR$$

where, k_{B} is Boltzmann's constant in joules per kelvin, T is the resistor's absolute temperature in kelvins, and R is the resistor value in ohms (Ω). Use this equation for quick calculation, at room temperature:

$$\sqrt{\overline{v_n^2}} = 0.13\sqrt{R}\,\mathrm{nV}/\sqrt{\mathrm{Hz}}.$$

For example, a 1 kΩ resistor at a temperature of 300 K has,

$$\sqrt{\overline{v_n^2}} = \sqrt{4 \cdot 1.38 \cdot 10^{-23}\,\mathrm{J/K} \cdot 300\,\mathrm{K} \cdot 1000\,\Omega} = 4.07 \cdot 10^{-9}\,\mathrm{V}/\sqrt{\mathrm{Hz}}.$$

For a given bandwidth, the root mean square (RMS) of the voltage, v_n, is given by,

$$v_n = \sqrt{\overline{v_n^2}}\sqrt{\Delta f} = \sqrt{4k_{\mathrm{B}}TR\Delta f}$$

where, Δf is the bandwidth in hertz over which the noise is measured. For a 1 kΩ resistor at room temperature and a 10 kHz bandwidth, the RMS noise voltage is 400 nV. A useful rule of thumb to remember is that 50 Ω at 1 Hz bandwidth correspond to 1 nV noise at room temperature.

A resistor in a short circuit dissipates a noise power of,

$$P = v_n^2 / R = 4k_{\mathrm{B}}T\Delta f.$$

The noise generated at the resistor can transfer to the remaining circuit; the maximum noise power transfer happens with impedance matching when the Thévenin equivalent resistance of the remaining circuit is equal to the noise generating resistance. In this case each one of the two participating resistors dissipates noise in both itself and in the other resistor. Since only half of the source voltage drops across any one of these resistors, the resulting noise power is given by,

$$P = k_B T\Delta f$$

where P is the thermal noise power in watts. Notice that this is independent of the noise generating resistance.

Noise Current

The noise source can also be modeled by a current source in parallel with the resistor by taking the Norton equivalent that corresponds simply to dividing by R. This gives the root mean square value of the current source as:

$$i_n = \sqrt{\frac{4k_B T \Delta f}{R}}.$$

Thermal noise is intrinsic to all resistors and is not a sign of poor design or manufacture, although resistors may also have excess noise.

Noise Power in Decibels

Signal power is often measured in dBm (decibels relative to 1 milliwatt). From the equation above, noise power in a resistor at room temperature, in dBm, is then:

$$P_{\mathrm{dBm}} = 10\log_{10}(k_{\mathrm{B}} T \Delta f \times 1000)$$

where the factor of 1000 is present because the power is given in milliwatts, rather than watts. This equation can be simplified by separating the constant parts from the bandwidth:

$$P_{\mathrm{dBm}} = 10\log_{10}(k_{\mathrm{B}} T \times 1000) + 10\log_{10}(\Delta f)$$

which is more commonly seen approximated for room temperature (T = 300 K) as:

$$P_{\mathrm{dBm}} = -174 + 10\log_{10}(\Delta f)$$

where Δf is given in Hz; e.g., for a noise bandwidth of 40 MHz, Δf is 40,000,000.

Using this equation, noise power for different bandwidths is simple to calculate:

Bandwidth Δf	Thermal noise power	Notes
1 Hz	−174 dBm	Deep space noise is considered -198 dBm/(Hz·K)
10 Hz	−164 dBm	
100 Hz	−154 dBm	
1 kHz	−144 dBm	
10 kHz	−134 dBm	FM channel of 2-way radio
100 kHz	−124 dBm	
180 kHz	−121.45 dBm	One LTE resource block
200 kHz	−121 dBm	GSM channel
1 MHz	−114 dBm	Bluetooth channel

2 MHz	−111 dBm	Commercial GPS channel
3.84 MHz	−108 dBm	UMTS channel
6 MHz	−106 dBm	Analog television channel
20 MHz	−101 dBm	WLAN 802.11 channel
40 MHz	−98 dBm	WLAN 802.11n 40 MHz channel
80 MHz	−95 dBm	WLAN 802.11ac 80 MHz channel
160 MHz	−92 dBm	WLAN 802.11ac 160 MHz channel
1 GHz	−84 dBm	UWB channel

Thermal Noise on Capacitors

Thermal noise on capacitors is referred to as *kTC* noise, after the formula for the mean squared charge noise. Thermal noise in an RC circuit has an unusually simple expression, as the value of the resistance (*R*) drops out of the equation. This is because higher *R* contributes to more filtering as well as to more noise. The noise bandwidth of the RC circuit is 1/(4*RC*), which can substituted into the above formula to eliminate *R*. The mean-square and RMS noise voltage generated in such a filter are:

$$\overline{v_n^2} = k_B T / C$$

$$v_n = \sqrt{k_B T / C}.$$

Thermal noise in the resistor accounts for 100% of *kTC* noise, for any resistance value. Therefore, if the resistance and the capacitor are at different temperatures, the temperature of the resistor alone should be used in the above calculation.

In the extreme case of the *reset noise* left on a capacitor by opening an ideal switch, the resistance is infinite, yet the formula still applies; however, now the RMS must be interpreted not as a time average, but as an average over many such reset events, since the voltage is constant when the bandwidth is zero. In this sense, the Johnson noise of an RC circuit can be seen to be inherent, an effect of the thermodynamic distribution of the number of electrons on the capacitor, even without the involvement of a resistor.

The noise is not caused by the capacitor itself, but by the thermodynamic fluctuations of the amount of charge on the capacitor. Once the capacitor is disconnected from a conducting circuit, the thermodynamic fluctuation is *frozen* at a random value with standard deviation as given.

The reset noise of capacitive sensors is often a limiting noise source, for example in image sensors. As an alternative to the voltage noise, the reset noise on the capacitor can also be quantified as the electrical charge standard deviation, as,

$$Q_n = \sqrt{k_B T C}.$$

Since the charge variance is $k_B TC$, this noise is often called *kTC noise.*

Any system in thermal equilibrium has state variables with a mean energy of $kT/2$ per degree of freedom. Using the formula for energy on a capacitor ($E = ½CV^2$), mean noise energy on a capacitor can be seen to also be $½C(kT/C)$, or also $kT/2$. Thermal noise on a capacitor can be derived from this relationship, without consideration of resistance.

The *kTC* noise is the dominant noise source at small capacitors.

Noise of capacitors at 300 K		
Capacitance	$\sqrt{k_B T/C}$	Electrons
1 fF	2 mV	12.5 e⁻
10 fF	640 µV	40 e⁻
100 fF	200 µV	125 e⁻
1 pF	64 µV	400 e⁻
10 pF	20 µV	1250 e⁻
100 pF	6.4 µV	4000 e⁻
1 nF	2 µV	12500 e⁻

Generalized Forms

The $4k_B TR$ voltage noise described above is a special case for a purely resistive component for low frequencies. In general, the thermal electrical noise continues to be related to resistive response in many more generalized electrical cases, as a consequence of the fluctuation-dissipation theorem. A variety of generalizations are noted. All of these generalizations share a common limitation, that they only apply in cases where the electrical component under consideration is purely passive and linear.

Reactive Impedances

Nyquist's original paper also provided the generalized noise for components having partly reactive response, e.g., sources that contain capacitors or inductors. Such a component can be described by a frequency-dependent complex electrical impedance $Z(f)$. The formula for the power spectral density of the series noise voltage is,

$$S_{v_n v_n}(f) = 4k_B T\eta(f)\mathrm{Re}[Z(f)].$$

The function $\eta(f)$ is simply equal to 1 except at very high frequencies, or near absolute zero.

The real part of impedance, $\mathrm{Re}[Z(f)]$, is in general frequency dependent and so the

Johnson–Nyquist noise is not white noise. The rms noise voltage over a span of frequencies f_1 to f_2 can be found by integration of the power spectral density:

$$\sqrt{\langle v_n^2 \rangle} = \sqrt{\int_{f_1}^{f_2} S_{v_n v_n}(f) df} \, .$$

Alternatively, a parallel noise current can be used to describe Johnson noise, its power spectral density being,

$$S_{i_n i_n}(f) = 4k_{\mathrm{B}} T \eta(f) \mathrm{Re}[Y(f)]$$

where, $Y(f) = 1/Z(f)$ is the electrical admittance; note that,

$$\mathrm{Re}[Y(f)] = \mathrm{Re}[Z(f)] / |Z(f)|^2 .$$

Quantum effects at High Frequencies or Low Temperatures

Nyquist also pointed out that quantum effects occur for very high frequencies or very low temperatures near absolute zero. The function $\eta(f)$ is in general given by,

$$\eta(f) = \frac{hf / k_{\mathrm{B}} T}{e^{hf/k_{\mathrm{B}} T} - 1},$$

where, h is Planck's constant.

At very high frequencies $f \gtrsim k_B T / h$, the function $\eta(f)$ starts to exponentially decrease to zero. At room temperature this transition occurs in the terahertz, far beyond the capabilities of conventional electronics, and so it is valid to set $\eta(f) = 1$ for conventional electronics work.

Relation to Planck's Law

Nyquist's formula is essentially the same as that derived by Planck in 1901 for electromagnetic radiation of a blackbody in one dimension—i.e., it is the one-dimensional version of Planck's law of blackbody radiation. In other words, a hot resistor will create electromagnetic waves on a transmission line just as a hot object will create electromagnetic waves in free space.

In 1946, Dicke elaborated on the relationship, and further connected it to properties of antennas, particularly the fact that the average antenna aperture over all different directions cannot be larger than $\lambda^2 / (4\pi)$, where λ is wavelength. This comes from the different frequency dependence of 3D versus 1D Planck's law.

Multiport Electrical Networks

Richard Q. Twiss extended Nyquist's formulas to multi-port passive electrical networks, including non-reciprocal devices such as circulators and isolators. Thermal noise appears at every port, and can be described as random series voltage sources in series with each port. The random voltages at different ports may be correlated, and their amplitudes and correlations are fully described by a set of cross-spectral density functions relating the different noise voltages,

$$S_{v_m v_n}(f) = 2k_B T\eta(f)(Z_{mn}(f) + Z_{nm}(f)^*)$$

where the Z_{mn} are the elements of the impedance matrix Z. Again, an alternative description of the noise is instead in terms of parallel current sources applied at each port. Their cross-spectral density is given by,

$$S_{i_m i_n}(f) = 2k_B T\eta(f)(Y_{mn}(f) + Y_{nm}(f)^*)$$

where, $Y = Z^{-1}$ is the admittance matrix.

Continuous Electrodynamic Media

The full generalization of Nyquist noise is found in fluctuation electrodynamics, which describes the noise current density inside continuous media with dissipative response in a continuous response function such as dielectric permittivity or magnetic permeability. The equations of fluctuation electrodynamics provide a common framework for describing both Johnson–Nyquist noise and free-space blackbody radiation.

GRAY NOISE

Gray noise is a specific kind of spectral sound that is designed for the psychoacoustics of the human ear. It differs in frequency allocation from other kinds of noise that work differently across different frequencies. Some new technologies may use gray noise to achieve a certain type of distributed sound efficiently, in other words, at lower decibel levels.

Experts describe gray noise as a noise that has more power at the top of the frequency spectrum and at the bottom of the spectrum, with less density in the middle of the spectrum. Scientists point out that this complements the acoustics of the human ear, so that to a listener, it sounds like every frequency is equally loud. Others have described gray noise as sounding like a three-dimensional kind of sound, or a sound that is not flat, but has some depth to it. Some contrast gray noise to other kinds of noise, like white noise, which is widely used for sleeping aids.

Gray noise has also been useful in testing a person's hearing.

References

- Shot-noise-what-is, electronic-rf-noise, basic-concepts: electronics-notes.com, Retrieved 31 March, 2019

- Philippe cattin (2012-04-24). "image restoration: introduction to signal and image processing". Miac, university of basel. Retrieved 11 october 2013
- Burst-noise-what-is, electronic-rf-noise, basic-concepts: electronics-notes.com, Retrieved 14 July, 2019
- Tudor barbu (2013). "variational image denoising approach with diffusion porous media flow". Abstract and applied analysis. 2013: 8. Doi:10.1155/2013/856876
- John r. Barry; edward a. Lee; david g. Messerschmitt (2004). Digital communications. Sprinter. P. 69. Isbn 9780792375487
- Gray-noise, definition: techopedia.com, Retrieved 17 May, 2019

3

Types of Noise Pollution

On the basis of the noise type and its origin, noise pollution can be classified into various types. Some of them are roadway noise, occupational noise, aircraft noise pollution, loud music, environmental noise, train noise, etc. All these types of noise pollution have been carefully analyzed in this chapter.

ROADWAY NOISE

It is a common misconception that vehicle noise is largely the result of engine noise or even the "wind" noise of passing cars and trucks. In most instances, the dominant source—termed tire—pavement noise—is actually located where the rubber meets the road. Many countries have regulations in place that limit the overall noise from vehicles, and noise from tires in particular. However, it was not until recently that international government agencies have thought about the influence of roads. Although institutional inertia was arguably to blame, one legitimate obstacle was the lack of a relevant, accurate, and reproducible method to measure tire—pavement noise.

That technical limitation has been overcome with the adoption of techniques such as on-board sound intensity, a method developed at General Motors in the 1980s, refined under sponsorship of the California Department of Transportation, and standardized by the American Association of State Highway and Transportation Officials. Although the standard continues to evolve, nowadays the most common variant of OBSI employs two probes consisting of two microphones each. Each probe is positioned mere inches from the rolling tire and is capable of measuring the sound intensity—that is, the product of pressure and the acoustic velocity, which is the velocity of a small bit of air relative to the air as a whole. Because sound intensity is a vector quantity, intensity measurements provide useful information not derivable from scalar pressure values. The OBSI technique uses two measurement probes because the sounds emanating from the front of the tire—pavement contact patch are usually different from those emanating from the rear.

Thousands of tire pavement combinations have been evaluated with the OBSI method. To simplify the interpretation of the data, pavement engineers have adopted a standard reference test tire that allows different pavements to be more readily compared. That SRTT is standardized by ASTM International. Although not ideal for a number of reasons, the SRTT has a significant advantage in that it will be available for purchase for the foreseeable future. In contrast, most commercially available tires are subject to being discontinued as a result of market forces.

Sources of the Noise

The challenge of developing a measurement technique for tire—pavement noise has now been met. The nature of the sound generation, though, remains an enigma. For both tire and pavement, three principal characteristics govern sound generation: texture, porosity, and stiffness. The complications lie in the specific relationships between those characteristics and the sounds that result when they change.

The most significant characteristic is texture. When driving a vehicle, you may experience roads that are both rough and loud. Texture with a significant amplitude is to blame for the bad all-around experience. But the texture wavelengths responsible for a rough or a noisy ride are different. Long-wavelength, low-frequency texture causes a rough ride. Noise is mostly caused by short-wavelength, high-frequency texture.

Noise heard outside the vehicle is largely governed by different texture wavelengths than those responsible for the noise heard inside the vehicle. At highway speeds, texture with significant amplitudes and wavelengths of about 10-50 mm is largely to blame for noise that is readily radiated away from the vehicle; texture with wavelengths of 20-200 mm is responsible for in-vehicle noise. In both cases, the dominant noise-producing mechanism is termed impact. As the tire rolls over the pavement, displacements in the road surface or tire-tread pattern create vibrations in the tire structure. Those vibrations, in turn, lead to acoustical energy being radiated from the tire-tread elements and sidewall. The impact mechanism is most evident in the characteristic roar of “knobby” tires found on some off-road or heavy commercial vehicles.

For one thing, the road would not be safe; absent texture, friction would be low and braking distance could be compromised. Even in terms of noise, engineers have learned that a perfectly flat road is not a quiet road. The reasons lie in the mechanics of the tire—pavement interaction. The realization that texture can reduce tire—pavement noise illustrates the complexity of the relationship between texture and noise.

The quest for a quieter pavement is often complicated by another broad category of texture—noise mechanisms termed aerodynamic. Those mechanisms are both numerous and complex, and their manifestation depends on the particular interaction of tire and pavement. As tire and pavement come into contact, so-called void geometries form in both the tire-tread blocks and the grooves and cavities in the pavement surface. Those geometries can have inherent resonant frequencies that are easily excited by the deformation and vibration of the tire itself. As the tire rolls, air trapped in the voids can be rapidly ejected—and that produces noise.

Numerous other sources of noise exist in addition to the impact and aerodynamic mechanisms. Various sources of vibration will generate sound, including those related to friction between the tire rubber and the finest texture. The characteristic squealing sound you hear when you corner in a smoothly paved parking garage illustrates how a frictional mechanism can sometimes dominate overall tire—pavement noise.

Texture is by far the most significant contributor to tire—pavement noise, but porosity has an effect too. Examples of highly porous materials deployed to reduce sound are all around, quite possibly right above your head. In pavement, the geometry of the void spaces and the thickness of the porous surface layer can be tuned to attenuate at the acoustical frequencies at which most of the energy is generated. The acoustical absorption of the pavement not only reduces the noise at the tire—pavement interface but also provides attenuation of the sound as it propagates out to the community. An additional benefit of porous pavements is that they provide drainage or escape paths for air that would otherwise be ejected from cavities between the rolling tire and the pavement. The result can be less noise.

Tire stiffness is another significant factor influencing the overall level of road noise. All else being equal, softer tires will result in a quieter ride. That said, noise is but one property of the tire that must be optimized; others include durability, fuel economy, and handling. Pavements, too, have different stiffnesses depending on how they are constructed. Concrete is typically stiffer than asphalt, but both materials are much stiffer than a tire. In neither case is the material's stiffness conducive to noise reduction: Pavement stiffness appears to be significant for noise attenuation only when it approaches that of a tire. In that case, however, the pavement's durability will be compromised. Still, innovative pavements termed poroelastic road surfaces (PERSs), constructed of epoxy-bound rubber particles, are currently under evaluation. Their acoustical performance appears quite good, due in part to their very low stiffness.

OCCUPATIONAL NOISE

Occupational noise is the amount of acoustic energy received by an employee's auditory system when they are working in the industry. Occupational noise, or industrial noise, is often a term used in occupational safety and health, as sustained exposure can cause permanent hearing damage.

"Twenty-two million workers are exposed to potentially damaging noise at work each year. Last year, U.S. business paid more than $1.5 million in penalties for not protecting workers from noise."

Occupational noise is considered an occupational hazard traditionally linked to loud industries such as ship-building, mining, railroad work, welding, and construction, but can be present in any workplace where hazardous noise is present.

Regulation

In the United States, the National Institute for Occupational Safety and Health (NIOSH) and the Occupational Safety and Health Administration (OSHA) work together to provide standards and regulations for noise in the workplace.

National Institute for Occupational Safety and Health (NIOSH), Occupational Safety and Health Administration (OSHA), Mine Safety and Health Administration (MSHA), Federal Railroad Administration (FRA) have all set standards on hazardous occupational noise in their respective industries. Each industry is different, as workers' tasks and equipment differ, but most regulations agree that noise becomes hazardous when it exceeds 85 decibels, for an 8-hour time exposure (typical work shift). This relationship between allotted noise level and exposure time is known as an Exposure action value (EAV) or Permissible exposure limit (PEL). The EAV or PEL can be seen as equations which manipulate the allotted exposure time according to the intensity of the industrial noise. This equation works as an inverse, exponential, relationship. As the industrial noise intensity increases, the allotted exposure tie, to still remain safe, decreases.

The above calculations of PEL and EAV are based on measurements taken to determine the intensity of that particular industrial noise. A-weighted measurements are commonly used to determine noise levels that can cause harm to the human ear. There are also special exposure meters available that integrate noise over a period of time to give an L_{eq} value (equivalent sound pressure level), defined by standards.

Risks of Occupational Hearing Loss

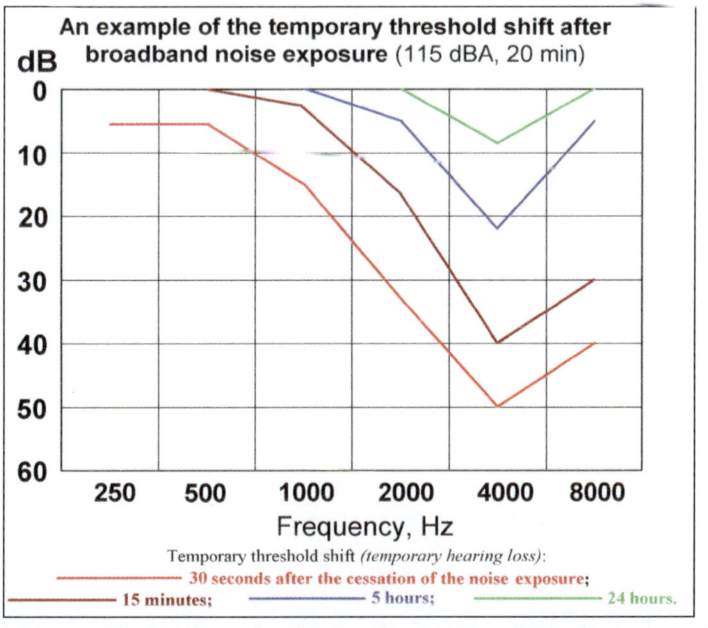

Temporary hearing loss after loud noise exposure. If such exposure is long enough, this temporary threshold shift may become permanent.

Occupational noise, if experienced repeatedly, at high intensity, for an extended period of time, can cause noise-induced hearing loss (NIHL) which is then classified as occupational hearing loss.

Noise, in the context of industrial noise, is hazardous to a person's hearing because of its loud intensity through repeated long-term exposure. In order for noise to cause hearing impairment for the worker, the noise has to be close enough, loud enough, and sustained long enough to damage the hair cells in the auditory system. These factors have been taken into account by the governing occupational health and safety organization to determine the unsafe noise exposure levels and durations for their respective industries.

Noise can also affect the safety of the employee and others. Noise can be a causal factor in work accidents as it may mask hazards and warning signals and impede concentration. High intensity noise interferes with vital workplace communication which increases the chance of accidents and decreases productivity.

Noise may also act synergistically with other hazards to increase the risk of harm to workers. In particular, toxic materials (e.g. some solvents, metals, asphyxiants and pesticides) have some ototoxic properties that may affect hearing function.

Modern thinking in occupational safety and health further identifies noise as hazardous to workers' safety and health. This hazard is experienced in various places of employment and through a variety of sources.

Noise, in the context of industrial noise, is hazardous to a persons hearing because of its loud intensity through repeated long-term exposure. In order for Noise to cause Hearing impairment for the worker, the noise has to be close enough, loud enough and the listener has to be exposed for long enough. These factors have been taken into account by the governing occupational health and safety organizations as they determine the unsafe noise exposure levels and durations for their respective industries.

National Institute for Occupational Safety and Health (NIOSH), Occupational Safety and Health Administration (OSHA), Mine Safety and Health Administration (MSHA), Federal Railroad Administration (FRA) have all set standards on hazardous occupational noise in their respective industries. Each industry is different, as workers tasks and equipment differ, but most regulations agree that noise becomes hazardous when it exceeds 85 decibels, for an 8-hour exposure (typical work shift). This relationship between allotted noise level and exposure time is known as an Exposure action value (EAV) or Permissible exposure limit (PEL). The EAV or PEL can be seen as equations which manipulate the allotted exposure time according to the intensity of the industrial noise. This equation works as an inverse relationship. As the industrial noise intensity increases, the allotted exposure time, to still remain safe, decreases.

Calculations of PEL and EAV are based on measurements taken to determine the intensity of that particular industrial noise. A-weighted measurements are commonly used to determine noise levels that can cause harm to the human ear. There are also special exposure meters available that integrate noise over a period of time to give an L_{eq} value (equivalent sound pressure level), defined by standards.

Reduction

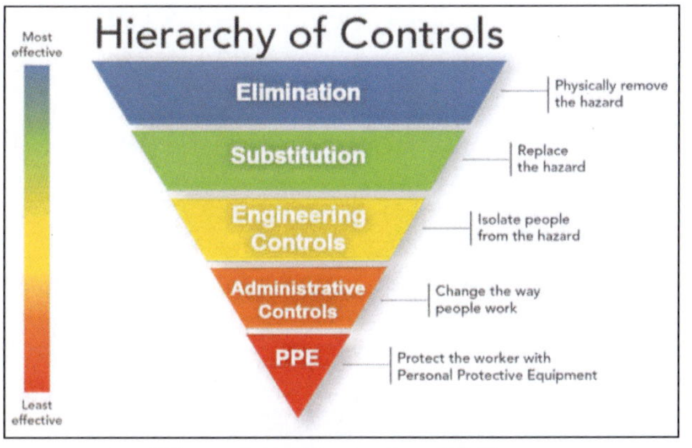

Hierarchy of Controls guideline.

There are several ways to limit your exposure to hazardous occupational noise. The hierarchy of controls is a guideline for reducing hazardous noise. First, the company can eliminate the noise source. If the noise source cannot be eliminated, the company must try to reduce the noise with alternative methods. This process is called acoustic quieting.

Acoustic quieting is the process of making machinery quieter by damping vibrations to prevent them from reaching the observer. The company can isolate the certain piece of machinery by placing materials on the machine or in between the machine and the worker to decreases the signal intensity that reaches the worker's ear.

To decrease an employee's exposure to hazardous noise, the company can also take administrative control by limiting the employee's exposure time. This can be done by changing work shifts and switching employees out from the noise exposure area. Lastly, to decrease occupational noise exposure, hearing protection should be used. There are several types of earplugs and earmuffs that can be used to attenuate the noise to a safe level.

AIRCRAFT NOISE POLLUTION

Aircraft noise pollution is a harmful noise effect produced by any aircraft or its components during the various phases of a flight. Sound production is divided into three categories:

- Mechanical noise: Rotation of the engine parts, most noticeable when fan blades reach supersonic speeds.

- Aerodynamic noise: From the airflow around the surfaces of the aircraft, especially when flying low at high speeds.
- Noise from aircraft systems: Cockpit and cabin pressurization and conditioning systems, and Auxiliary Power units.

Health consequences include sleep disturbance, hearing impairment and heart disease, as well as workplace accidents caused by stress. Memory and recall can also be affected. Governments have enacted extensive controls that apply to aircraft designers, manufacturers, and operators, resulting in improved procedures and cuts in pollution.

A Qantas Boeing 747-400 passes close to houses on the boundary of London Heathrow Airport, England.

Mechanisms of Sound Production

Small general aviation aircraft produce localized aircraft noise.

Aircraft noise is noise pollution produced by an aircraft or its components, whether on the ground while parked such as auxiliary power units, while taxiing, on run-up from propeller and jet exhaust, during take off, underneath and lateral to departure and arrival paths, over-flying while en route, or during landing. A moving aircraft including the jet engine or propeller causes compression and rarefaction of the air, producing motion of air molecules. This movement propagates through the air as pressure waves. If these pressure waves are strong enough and within the audible frequency spectrum, a sensation of hearing is produced. Different aircraft types have different noise levels and frequencies. The noise originates from three main sources:

- Engine and other mechanical noise;
- Aerodynamic noise;
- Noise from aircraft systems.

Helicopter main and tail rotors produce aerodynamic noise.

Engine and other Mechanical Noise

NASA researchers at Glenn Research Center conducting tests on jet engine noise in 1967.

Much of the noise in propeller aircraft comes equally from the propellers and aerodynamics. Helicopter noise is aerodynamically induced noise from the main and tail rotors and mechanically induced noise from the main gearbox and various transmission chains. The mechanical sources produce narrow band high intensity peaks relating to the rotational speed and movement of the moving parts. In computer modelling terms noise from a moving aircraft can be treated as a line source.

Aircraft gas turbine engines (jet engines) are responsible for much of the aircraft noise during takeoff and climb, such as the *buzzsaw noise* generated when the tips of the fan blades reach supersonic speeds. However, with advances in noise reduction technologies—the airframe is typically more noisy during landing.

The majority of engine noise is due to jet noise—although high bypass-ratio turbofans do have considerable fan noise. The high velocity jet leaving the back of the engine has an inherent shear layer instability (if not thick enough) and rolls up into ring vortices. This later breaks down into turbulence. The SPL associated with engine noise is proportional to the jet speed (to a high power). Therefore, even modest reductions in exhaust velocity will produce a large reduction in jet noise.

Engines are the main source of aircraft noise. The geared Pratt & Whitney PW1000G helped reduce the noise levels of the Bombardier CSeries, Mitsubishi MRJ and Embraer E-Jet E2 crossover narrowbody aircraft: the gearbox allows the fan to spin at an optimal speed, which is one third the speed of the LP turbine, for slower fan tip speeds. It has a 75% smaller noise footprint than current equivalents. The PowerJet SaM146 in the Sukhoi Superjet 100 features 3D aerodynamic fan blades and a nacelle with a long mixed duct flow nozzle to reduce noise.

Aerodynamic Noise

Aerodynamic noise arises from the airflow around the aircraft fuselage and control surfaces. This type of noise increases with aircraft speed and also at low altitudes due to the density of the air. Jet-powered aircraft create intense noise from aerodynamics. Low-flying, high-speed military aircraft produce especially loud aerodynamic noise.

The shape of the nose, windshield or canopy of an aircraft affects the sound produced. Much of the noise of a propeller aircraft is of aerodynamic origin due to the flow of air around the blades. The helicopter main and tail rotors also give rise to aerodynamic noise. This type of aerodynamic noise is mostly low frequency determined by the rotor speed.

Typically noise is generated when flow passes an object on the aircraft, for example, the wings or landing gear. There are broadly two main types of airframe noise:

- Bluff Body Noise: The alternating vortex shedding from either side of a bluff body, creates low-pressure regions (at the core of the shed vortices) which manifest themselves as pressure waves (or sound). The separated flow around the bluff body is quite unstable, and the flow “rolls up” into ring vortices—which later break down into turbulence.

- Edge Noise: When turbulent flow passes the end of an object or gaps in a structure (high lift device clearance gaps) the associated fluctuations in pressure are heard as the sound propagates from the edge of the object (radially downwards).

Noise from Aircraft Systems

Cockpit and cabin pressurization and conditioning systems are often a major contributor within cabins of both civilian and military aircraft. However, one of the most significant sources of cabin noise from commercial jet aircraft, other than the engines, is the Auxiliary Power Unit (APU), an onboard generator used in aircraft to start the main engines, usually with compressed air, and to provide electrical power while the aircraft is on the ground. Other internal aircraft systems can also contribute, such as specialized electronic equipment in some military aircraft.

Health Effects

Aircraft engines are the major source of noise and can exceed 140 decibels (dB) during takeoff. While airborne, the main sources of noise are the engines and the high speed turbulence over the fuselage.

There are health consequences of elevated sound levels. Elevated workplace or other noise can cause hearing impairment, hypertension, ischemic heart disease, annoyance, sleep disturbance, and decreased school performance. Although some hearing loss occurs naturally with age, in many developed nations the impact of noise is sufficient to impair hearing over the course of a lifetime. Elevated noise levels can create stress, increase workplace accident rates, and stimulate aggression and other anti-social behaviors. Airport noise has been linked to high blood pressure.

FAA Advice

The Federal Aviation Administration (FAA) regulates the maximum noise level that individual civil aircraft can emit through requiring aircraft to meet certain noise certification standards. These standards designate changes in maximum noise level requirements by “stage” designation. The U.S. noise standards are defined in the Code of Federal Regulations (CFR) Title 14 Part 36 – Noise Standards: Aircraft Type and Airworthiness Certification (14 CFR Part 36). The FAA says that a maximum day-night average sound level of 65 dB is incompatible with residential communities. Communities in affected areas may be eligible for mitigation such as soundproofing.

Cabin Noise

Aircraft noise also affects people within the aircraft: crew and passengers. Cabin noise can be studied to address the occupational exposure and the health and safety of pilots and flight attendants. In 1998, 64 commercial airline pilots were surveyed regarding hearing loss and tinnitus. In 1999, the NIOSH conducted several noise surveys and health hazard evaluations, and found noise levels exceeding its recommended exposure limit of 85 A-weighted decibels as an 8-hr TWA. In 2006, the noise levels inside an Airbus A321 during cruise have been reported as approximately 78 dB(A) and during taxi when the aircraft engines are producing minimal thrust, noise levels in the cabin have been recorded at 65 dB(A). In 2008, a study of Swedish airlines cabin crews found average sound levels between 78–84 dB(A) with maximum A-weighted exposure of 114 dB but found no major hearing threshold shifts. In 2018, a study of sound levels measured on 200 flights representing six aircraft groups found media noise level of 83.5 db(A) with levels reaching 110 dB(A) on certain flights, but only 4.5% exceeded the NIOSH recommended 8-hr TWA of 85 dB(A).

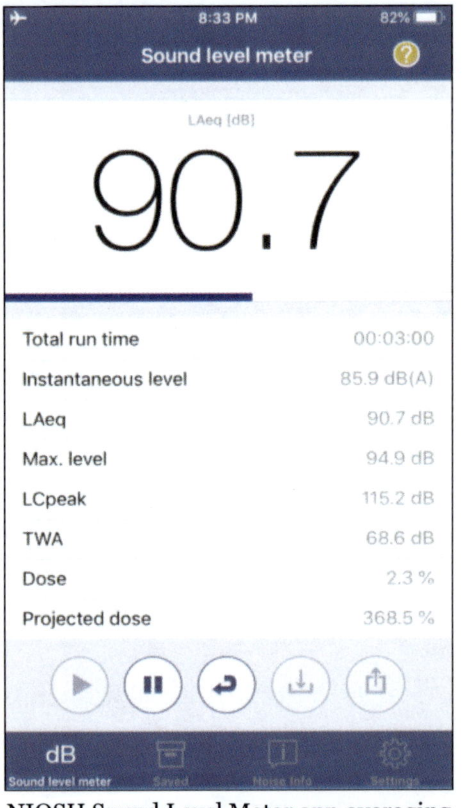

NIOSH Sound Level Meter app averaging 90.7 dB at takeoff in a Boeing 777.

Cognitive Effects

Simulated aircraft noise at 65 dB(A) has been shown to negatively affect individuals' memory and recall of auditory information. In one study comparing the effect of aircraft noise to the effect of alcohol on cognitive performance, it was found that simulated aircraft noise at 65 dB(A) had the same effect on individuals' ability to recall auditory information as being intoxicated with a Blood Alcohol Concentration (BAC) level of at 0.10. A BAC of 0.10 is double the legal limit required to operate a motor vehicle in many developed countries such as Australia.

LOUD MUSIC

Loud music is music that is played at a high volume, often to the point where it disturbs others and/or causes hearing damage. It may include music that is sung live with one or more voices, played with instruments, or broadcast with electronic media, such as radio, CD, or MP3 players.

Music that is played at an extremely high volume can be annoying to others, is considered by many to be disrespectful, and can sometimes be unlawful.

Playing loud music that can be heard from outside of the property from where it is being played (such as a house, apartment, hotel room, or motor vehicle) is considered to be rude by many people and societies. Among those opposed to the practice, it may result in the loss of respect and possible legal action. But in certain contained settings, such as clubs or concerts, music is often played very loudly, but is viewed as acceptable.

Consequences

Criminal

Many jurisdictions have laws defining loud music as a criminal offense, typically a misdemeanor. The exact definition of what constitutes a loud music violation varies by location, either at a certain volume (measured in decibels) or the distance from the source at which the music can be heard. The time of day is also often a factor in the law, with the restrictions in some places applying only to specified nighttime hours (e.g. 11 PM-7 AM). The amount of effort put forth by law enforcement members in dealing with loud music also varies by location.

The most common punishment for a conviction is a fine or some other small sanction. But on rare occasions, loud music may be grounds for imprisonment. In May 2008, a

United Kingdom woman was sentenced to 90 days in jail for violating a court order not to play music that disturbed her neighbors eleven times.

Police have also at times discovered other crimes, such as illegal drug usage, when investigating loud music complaints.

Many public transportation services have rules against the use of sound-producing devices without earphones, or even with earphones if the music can be heard by others. Since mass transit agencies are frequently government-operated and/or subsidized, these rules can be legally enforced, and violation may result in prosecution.

Civil

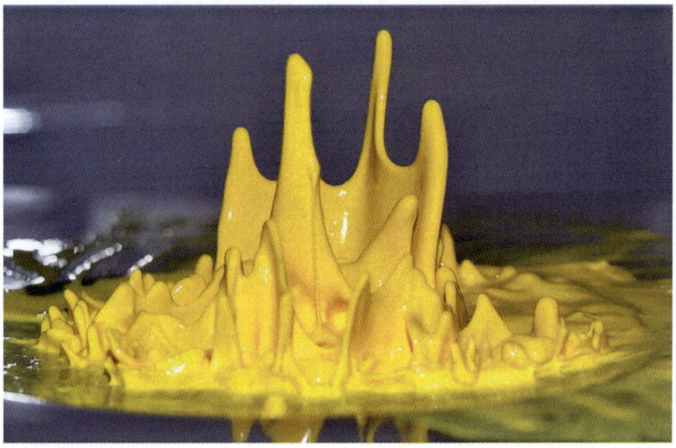

Yellow paint on a piece of film wrap over a bass speaker playing very loud music.

In many settings, loud music is not tolerated by property owners, and may be grounds for certain civil actions, such as eviction from rented property.

Property owners at locations where patrons visit temporarily, such as hotels, campgrounds, or businesses, may order those who play loud music to leave the property.

Health

Continual exposure to loud music may result in hearing loss. The National Institute for Occupational Safety and Health (NIOSH) has developed a set of recommendations aimed at protecting the hearing health of musicians and those who work in music and entertainment venues. Depending on the music sound levels and duration of exposure, as well as hearing protection used if any, the risk of hearing damage can vary significantly. Music played at 85 decibels for prolonged periods of times can cause hearing damage, for instance, sound levels at some rock concerts can reach 110-120 A-weighted decibels, and at those levels, the maximum daily limit set by most standards and regulations can be reached in less than one minute of exposure.

Continual exposure to loud music can also lead to tinnitus.

It is predicted that exposure to loud music will cause as many as 50 million Americans to suffer hearing loss by 2050.

Excessive Drinking

A study conducted by French scientists showed that loud music leads to more alcohol consumption in less time. For three Saturday evenings researchers observed customers of two bars situated in a medium-sized city in the west of France. Participants included forty males aged between 18 and 25, who were unaware that they were subjects of a research. The study featured only those who ordered a glass of draft beer (25 cl. or 8 oz.). The lead researcher, Nicolas Guéguen, said that each year more than 70,000 people in France die from an increased level of alcohol consumption, which also leads to fatal car accidents.

LOMBARD EFFECT

The Lombard effect or Lombard reflex is the involuntary tendency of speakers to increase their vocal effort when speaking in loud noise to enhance the audibility of their voice. This change includes not only loudness but also other acoustic features such as pitch, rate, and duration of syllables. This compensation effect results in an increase in the auditory signal-to-noise ratio of the speaker's spoken words.

Great tits sing at a higher frequency in noise polluted urban surroundings than quieter ones to help overcome the auditory masking that would otherwise impair other birds hearing their song. Although great tits achieve a change in song frequency by switching song types, in other urban birds the change in frequency might be related to the Lombard effect. For instance, in humans, the Lombard effect results in speakers adjusting frequency.

The effect links to the needs of effective communication, as there is a reduced effect when words are repeated or lists are read where communication intelligibility is not important. Since the effect is involuntary it is used as a means to detect malingering in those simulating hearing loss. Research on birds and monkeys find that the effect also occurs in the vocalizations of animals.

The effect was discovered in 1909 by Étienne Lombard, a French otolaryngologist.

Lombard Speech

Listeners hear a speech recorded with background noise better than they hear a speech which has been recorded in quiet with masking noise applied afterwards. This is because changes between normal and Lombard speech include:

- Increase in phonetic fundamental frequencies;
- Shift in energy from low frequency bands to middle or high bands;
- Increase in sound intensity;
- Increase in vowel duration;
- Spectral tilting;
- Shift in formant center frequencies for F1 (mainly) and F2;
- The duration of content words are prolonged to a greater degree in noise than function words;
- Great lung volumes are used;
- It is accompanied by larger facial movements, though these do not aid as much as sound changes.

These changes cannot be controlled by instructing a person to speak as they would in silence, though people can learn control with feedback.

The Lombard effect also occurs following laryngectomy when people following speech therapy talk with esophageal speech.

Mechanisms

The intelligibility of an individual's own vocalization can be adjusted with audio-vocal reflexes using their own hearing (private loop), or it can be adjusted indirectly in terms of how well listeners can hear the vocalization (public loop). Both processes are involved in the Lombard effect.

Private Loop

A speaker can regulate their vocalizations, particularly their amplitude relative to background noise, with reflexive auditory feedback. Such auditory feedback is known to maintain the production of vocalization since deafness affects the vocal acoustics of both humans and songbirds. Changing the auditory feedback also changes vocalization in human speech or bird song. Neural circuits have been found in the brainstem that enable such reflex adjustment.

Public Loop

A speaker can regulate their vocalizations at higher cognitive level in terms of observing its consequences on their audience's ability to hear it. In this auditory self-monitoring adjusts vocalizations in terms of learnt associations of what features of their vocalization, when made in noise, create effective and efficient communication. The Lombard effect has been found to be greatest upon those words that are important to the listener to understand a speaker suggesting such cognitive effects are important.

Development

Both private and public loop processes exist in children. There is a development shift however from the Lombard effect being linked to acoustic self-monitoring in young children to the adjustment of vocalizations to aid its intelligibility for others in adults.

Neurology

The Lombard effect depends upon audio-vocal neurons in the periolivary region of the superior olivary complex and the adjacent pontine reticular formation. It has been suggested that the Lombard effect might also involve the higher cortical areas that control these lower brainstem areas.

Choral Singing

Choral singers experience reduced feedback due to the sound of other singers upon their own voice. This results in a tendency for people in choruses to sing at a louder level if it is not controlled by a conductor. Trained soloists can control this effect but it has been suggested that after a concert they might speak more loudly in noisy surroundings, such as after-concert parties.

The Lombard effect also occurs to those playing instruments such as the guitar.

Animal Vocalization

Noise has been found to affect the vocalizations of animals that vocalize against a background of human noise pollution. Experimentally, the Lombard effect has also been found in the vocalization of:

- Budgerigars
- Canaries

- Cats
- Chickens
- Common marmosets
- Cottontop tamarins
- Japanese quail
- Nightingales
- Rhesus macaques
- Squirrel monkey
- Zebra finches
- Beluga whales
- Pale spear-nosed bats

ENVIRONMENTAL NOISE

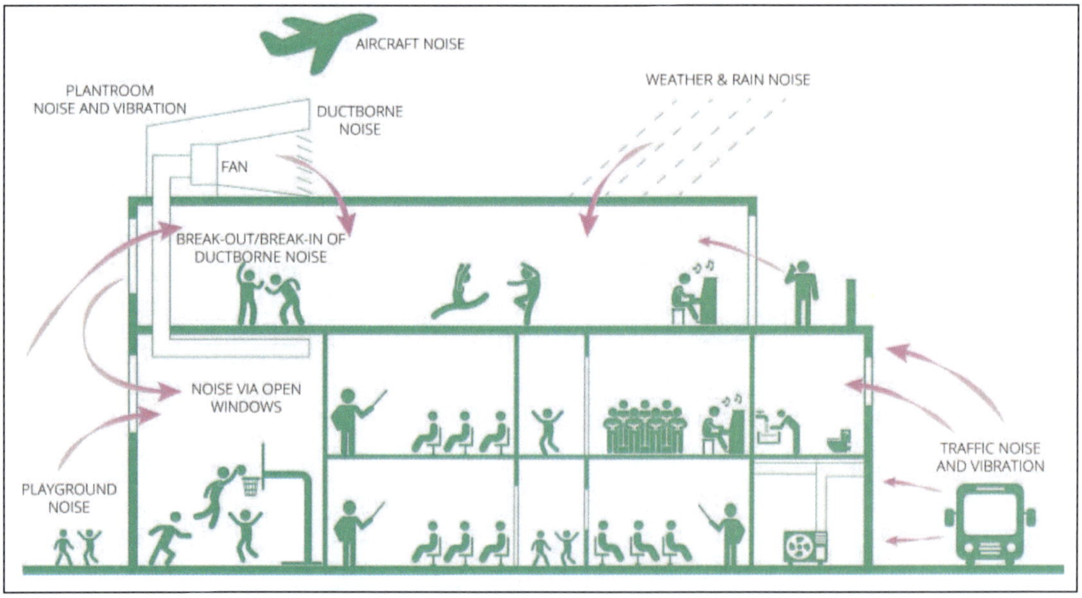

Example of transportation noise and how it affects the community.

Environmental noise is an accumulation of noise pollution that occurs outside. This noise can be caused by transport, industrial, and recreational activities.

Noise is frequently described as 'unwanted sound'. Within this context, environmental noise is generally present in some form in all areas of human, animal, or environmental activity. The effects in humans of exposure to environmental noise may vary from emotional to physiological and psychological.

Noise at low levels is not necessarily harmful. Environmental noise can also convey a sense of liveliness in an area, which can be desirable. However, the adverse effects of noise exposure (i.e. noise pollution) could include: interference with speech or other 'desired' sounds, annoyance, sleep disturbance, anxiety, hearing damage and stress-related cardiovascular health problems.

As a result, environmental noise is studied, regulated, and monitored by many governments and institutions around the world. This creates a number of different occupations. The basis of all decisions is supported by the objective and accurate measurement of noise. Noise is measured in decibels (dB) using a pattern-approved sound level meter. The measurements are typically taken over a period of weeks, in all weather conditions.

Environmental Noise Emission

Noise from transportation is typically emitted by machinery (e.g. the engine or exhaust) and aerodynamic noise caused by the compression and friction in the air around the vessel during motion. Environmental noise from the railway specifically is variable depending on the speed and quality of the tracks used for transportation.

Industrial and recreational noise could be generated by a multitude of different sources and processes. Industrial noise can be generated by factories and plants (i.e., product fabrication or assembly), power generation (hydroelectricity or wind turbines), construction activities, or agricultural and meat processing facilities. Sources of recreational noise vary widely but they can include music festivals, shooting ranges, sporting events, car racing, woodworking, pubs, people's activities on the street, etc.

Sound propagation outdoors is subject to meteorological effects (e.g. wind, temperature) that affect the distance, speed, and direction with which environmental noise travels from a source to a listener.

Environmental Noise and Health Effects

Noise and quality of life are correlated. The increase of environmental noise, especially for those living near railways and airports, has created conflict. Getting adequate and quality sleep is difficult for those who live in areas of high noise exposure. When the body is at rest, noise stimuli is continually being presented in the environment. The body responds to these sounds which can negatively affect sleep.

High exposure to environmental noise can play a role in cardiovascular disease. Noise can raise blood pressure, change heart rate, and release stress hormones. Consistent changes in these health stats can lead to risks for hypertension, arteriosclerosis, and even more serious events such as a stroke or myocardial infarction.

Environmental Noise Policy and Regulation

United States

The Noise Control Act of 1972 established a U.S. national policy to promote an environment for all Americans to be free from noise that jeopardizes their health and welfare. In the past, Environmental Protection Agency (EPA) coordinated all federal noise control activities through its Office of Noise Abatement and Control. The EPA phased out the office's funding in 1982 as part of a shift in federal noise control policy to transfer the primary responsibility of regulating noise to state and local governments. However, the Noise Control Act of 1972 and the Quiet Communities Act of 1978 were never rescinded by Congress and remain in effect today, although essentially unfunded.

Today, in the absence of a national guidance and enforcement by the EPA, states, cities, and municipalities have had little or no guidance on writing competent and effective noise regulations. Since the EPA last published its Model Community Noise Ordinance in 1974, communities have struggled to develop their ordinances, often relying on copying guidance from other communities, and sometimes copying their mistakes. Noise laws and ordinances vary widely among municipalities though most specify some general prohibition against making noise that is a nuisance and the allowable sound levels that can cross a property line. Some ordinances set out specific guidelines for the level of noise allowable at certain times of the day and for certain activities.

The Federal Aviation Administration (FAA) regulates aircraft noise by specifying the maximum noise level that individual civil aircraft can emit through requiring aircraft to meet certain noise certification standards. These standards designate changes in maximum noise level requirements by "stage" designation. The U.S. noise standards are defined in the Code of Federal Regulations (CFR) Title 14 Part 36 – Noise Standards: Aircraft Type and Airworthiness Certification (14 CFR Part 36). The FAA also pursues a program of aircraft noise control in cooperation with the aviation community. The FAA has set up a process to report aviation-related noise complaints for anyone who may be impacted by Aircraft noise.

The Federal Highway Administration (FHWA) developed noise regulations to control highway noise as required by the Federal-Aid Highway Act of 1970. The regulations requires promulgation of traffic noise-level criteria for various land use activities, and describe procedures for the abatement of highway traffic noise and construction noise.

The Department of Transportation's Bureau of Transportation Statistics has created a National Transportation Noise Map to provide access to comprehensive aircraft and road noise data on national and county-level. The map aims to assist city planners, elected officials, scholars, and residents to gain access to up-to-date aviation and Interstate highway noise information.

European Union

The European Union has a special definition based on the European directive 2002/49/EC article 10.1. This directive gives a definition for environmental noise. The main goal is to create an integrated noise management system.The Environmental Noise Directive (END) was created in the European Union to provide guidelines, laws, and standards in the management of environmental noise. The END has created noise mapping, noise action plans, and quiet areas to control environmental noise and the negative effects it can have on individuals.

The implementation is divided into phases: In the first phase, the member states shall inform about major roads with more than six million vehicles a year, major railways with more than 60,000 trains per year, major airports with more than 50,000 movements per year and metropolitan areas with more than 250,000 inhabitants. In the second phase, these numbers are halved; only the criteria for airports remains unchanged. In the third and the following phases, the methods for calculation of the noise levels will change while the criteria remains unchanged. Each phase consists of three steps: the collection of the data from the main sources of noise, strategic noise maps and action plans. The countries listed follow the guidelines of the European Union.

Austria

In Austria the institution which is responsible for the noise sources is also responsible for the noise maps concerning these sources. This means that the Federation is responsible for the federal roads and each state is responsible for the country's roads.

France

France reported 24 metropolitan areas. Paris was the biggest with 9.6 million inhabitants and 272 square kilometres.

Germany

Germany implemented national regulations in 2005 and 2006 and reported 27 metropolitan areas in the first phase. Berlin was the most populated with 3.39 million inhabitants and 889 square kilometres, Hamburg was considered the largest with 1,045 square kilometres and 2 million inhabitants. The smallest was Gelsenkirchen with 270,000 inhabitants and 105 square kilometres. In the national legislation, noise resulting from recreational activities like sports and leisure is not considered as environmental noise.

United Kingdom

The United Kingdom reported a total of 28 metropolitan areas, where London is the largest with 8.3 million inhabitants. The majority of metropolitan areas are located in

England. There are two in Scotland and Wales, and one in Northern Ireland with it being Belfast, the country's capital.

TRAIN NOISE

The sound emitted from a train (locos and wagons) on the rail track during their operations is 'rail noise.' Noises created by a train may be heard inside the train and outside. Various parts of a train produce noise, and different kinds of train wheels produce different amounts of noise. Railway noise is also considered to be a type of environmental noise. There are a vast variety of sources and causes of rail noise. When a train is moving, several distinct sounds can be recognised such as the locomotive engine noise and the wheels' turning on the railroad track, braking squeal noise in curves, vibration from rail corrugation and out-ofround wheels, vehicle coupling in shunting yards, etc. The air displacement of a train or subway wagon in a tunnel can create different whooshing sounds. Trains also use horns, whistles, bells, and other noisemaking devices for both communication and warning. Trains propelled by electric traction motors controlled by high speed electronic inverters can produce a whining noise.

Factors of Rail Noise

There are two factors that determine railway noise, namely technical one, depending on the type of wagon, brake system used, weight of the wagon, speed with which it is moving, condition of the tracks (poor track maintenance can cause rough rails which increase noise level), and geographic one depending on landscape that can intensify the noise (resonance issue), or reduce it, closeness to the urban areas (freight trains are mostly operated at night time and they pass through highly populated areas) and infrastructure (noise from steel bridges, for instance).

Source of Rail Noise

Railway noise consists of various types of noise, namely rolling noise, traction and auxiliary noise, aerodynamic noise, curve squeal noise, impact noise, bridge noise etc. However, the rolling noise is considered to be the most predominant which increases with train speed (9 dB increase in sound level for doubling of speed). Whereas traction noise does not depend on train speed and it is dominant only at low speeds. Aerodynamic noise depends largely on the speed of the train and becomes dominant only at high speeds. Bridge noise or ground-born noise is mainly concentrated at low frequencies, while squeal noise occurs at very high frequencies.

Rolling Noise

Rolling noise is caused by wheel and rail vibrations induced at the wheel/rail interaction.

Roughness on the wheel and rail running surfaces induces vertical vibration of the wheel and rail systems according to their dynamic properties. The main wavelengths of roughness that are relevant to rolling noise are between about 5 and 500 mm. The vibration is transmitted into the wheel and track structures leading to sound radiation. Both wheel and track vibration are important for noise level and both must be taken into account. Rolling noise is broad-band in nature and when train increase speed high frequencies are of higher importance. It is also important to stress here that the vibrations of both wheel and rail are induced by the combination of their roughnesses. For instance, wheels fitted with cast-iron brake blocks have large roughness with wavelengths ca. 40-80 mm. When train moves with speed 100 km/h this roughness excites frequencies where the track vibration radiates most sound. In this case wheel and track are responsible for the noise. Impact of noise is excited by irregularities such as wheel flats, rail joints, dipped welds or switchers and crossing. It is similar to rolling noise as the excitation is primarily vertical, yet it differs in that non-linearity in the contact zone is important and cannot be neglected.

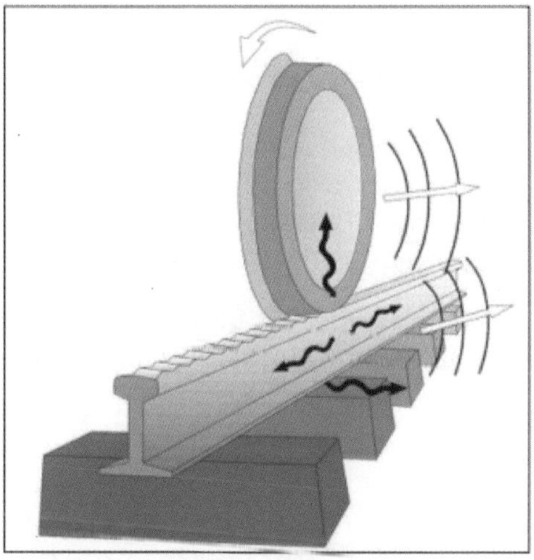

Another important observation that has to be mentioned regarding rolling noise is that the heavier a wagon is, the less noise it emits, for there is a better contact between wheels and rails. However, heavy wagons destroy the rail more quickly and damage the wheels. Therefore, there are limits set for maximum weights for axels and rail. Talking about the FVL wagons (H-type – closed wagons or L-type – open wagons), they are considered to be the lightest among all wagon classes with a wheel diameter of 680mm. With small diameter the wheels need to rotate quicker, which leads to an increase in friction and an increase of rolling noise. Furthermore, a smaller wheels have smaller contact with rail than bigger in diameter wheels. Therefore, the size of wheels also influences a noise level during their movement.

The FVL wagons are also believed to be the noisiest wagons, because of the contact between wheels and rails which is less strong due to their light weight than it is with other heavier types of wagons. It also needs to be mentioned here that the FVL wagons contain many small loose parts which are used either for lashing and securing wheels of cars, or fixing the decks of the wagons, etc. During the movement of these types of wagons when either loaded or empty, these loose parts emit lots of noise and it increases with the speed of a train. Therefore, the FVL wagons of both H- and L-types have additional sources of noise due to their technical specifications and design. They are believed to be the noisiest ones amongst all other types of wagons and additional solutions need to be proposed to tackle their noise level reduction.

Curve and Brake Squeal

This type of noise is caused by the interaction between wheel and rail but has a different character. It is strongly tonal and is associated with vibration of the wheel in one of its resonances. It is excited by unsteady transverse forces at the contact occurring during curving. Here distinction needs to be made between a squeal caused by lateral creepage, or 'top-of-rail' squeal, and 'flange squeal' or 'flanging noise'. The latter has a more intermittent nature and has higher frequency content. It consists of many more harmonics rather than a single dominant tone. Similar in tonal or multi-tonal noise is brake squeal which is emitted during braking.

Bridge Noise

When a train passes over a bridge the noise from the vibration of the bridge can increase considerably, depending on the type of bridge. The bridge is excited by dynamic forces acting on it from the track. Steel bridges with direct fastening are the noisiest and can be more than 10 dB noisier than plain ballasted track, in some case up to 20 dB. The increase in noise on a bridge is usually greatest at low frequency.

Aerodynamic Noise

Aerodynamic noise is caused by unsteady air flow over the train. This type of noise source increases rapidly with speed of a train. Both broad-band and tonal noise can be generated by air flow over various parts of the train, yet much of the sound energy is concentrated in the lower part of the frequency region. It was only in recent years that the source of aerodynamic noise was understood, but it is much more complicated to model.

Ground Vibration and Noise

Vibrations transmitted through the ground can be detected in two ways, namely low frequency vibration between about 2 and 80 Hz is perceived as feeble 'whole body' vibration, which is associated mainly with heavy freight trains in particular areas; higher frequency ground-born vibration from about 30 to 250 Hz cause the walls, floors and ceilings of a building to vibrate and radiate low frequency noise. This noise is more associated with a train passing through tunnels in urban areas. It can also be significant for surface railways, especially where noise barriers block out the direct airborne sound. In this case air is a uniform medium, whereas the ground can have a layered structure and its properties vary from one site to another. It is also difficult to predict absolute levels of vibration. It requires complex models of the ground as well as models of the train and track and the buildings at the receiver location.

Other Sources of Railway Noise

There are other sources of rail noise, such as traction noise from diesel engines, exhaust and intake, from traction motors and fans, from gearboxes, turbocharges, etc.; warning signals from trains (horns, etc.) and fixed installations (e.g. level crossings); track maintenance equipment; shunting noise, in particular the noise from impact between wagons and locos.

Internal Noise and Vibrations

There are many other sources of sounds that are relevant to environmental noise that lead to noise inside the railway vehicle. Their frequency content is modified by the transmission paths into the vehicle, which may be both structure-borne and air-borne. The noise spectrum inside tends to be dominated by low frequency sound. Furthermore, low frequency vibration inside the train affects the comfort of a journey.

References

- "Safety and health topics | occupational noise exposure | occupational safety and health administration". Www.osha.gov. Retrieved 2016-11-08
- izmerov, nikolai; suvorov, herman; prokopenko, ludmila (2001). "chapter 4. Occupational hearing loss". The man and the noise (человек и шум) (in russian). Moscow: геотар-мед. P. 103. Isbn 5-9231-0057-6
- "aircraft airframe noise—research overview". Archived from the originalon 2008-05-17. Retrieved 2008-07-13
- Slabbekoorn h, peet m (july 2003). "ecology: birds sing at a higher pitch in urban noise". Nature. 424 (6946): 267. Bibcode:2003natur.424..267s. Doi:10.1038/424267a. Pmid 12867967
- Kinsler, l.e., frey, a.r., coppens, a.b. And sanders, j.v. (2000). Fundamentals of acoustics. New york city: john wiley & sons. P. 359. Isbn 978-0471-84789-2

4

Effects of Noise Pollution

Noise has many harmful and disturbing effects on animals and humans. The major effects of noise pollution in humans include acoustic trauma and noise-induced hearing loss. This chapter has been carefully written to provide an easy understanding of the various adverse effects of noise on humans as well as wildlife.

EFFECT ON WILDLIFE

Wildlife faces far more problems than humans because of noise pollution since they are more dependent on sound. Animals develop a better sense of hearing than us since their survival depends on it. The ill effects of excessive noise begin at home. Pets react more aggressively in households where there is a constant noise.

They become disoriented more easily and face many behavioural problems. In nature, animals may suffer from hearing loss, which makes them easy prey and leads to dwindling populations. Others become inefficient at hunting, disturbing the balance of the eco-system.

Species that depend on mating calls to reproduce are often unable to hear these calls due to excessive man-made noise. As a result, they are unable to reproduce and cause declining populations. Others require sound waves to echolocate and find their way when migrating. Disturbing their sound signals means they get lost easily and do not migrate when they should. To cope up with the increasing sound around them, animals are becoming louder, which may further add to the pollution levels. This is why understanding noise pollution can help us lower the impact it has on the environment.

NOISE-INDUCED HEARING LOSS

Noise-induced hearing loss (NIHL) is hearing impairment resulting from exposure to loud sound. People may have a loss of perception of a narrow range of frequencies or impaired perception of sound including sensitivity to sound or ringing in the ears.

When exposure to hazards such as noise occur at work and is associated with hearing loss, it is referred to as occupational hearing loss.

Noise-induced hearing loss	

The international symbol of deafness and hearing loss.	
Specialty	Otorhinolaryngology, audiology

Hearing may deteriorate gradually from chronic and repeated noise exposure (such as to loud music or background noise) or suddenly from exposure to impulse noise, which is a short high intensity noise (such as a gunshot or airhorn). In both types, loud sound overstimulates delicate hearing cells, leading to the permanent injury or death of the cells. Once lost this way, hearing cannot be restored in humans.

There are a variety of prevention strategies available to avoid or reduce hearing loss. Lowering the volume of sound at its source, limiting the time of exposure and physical protection can reduce the impact of excessive noise. If not prevented, hearing loss can be managed through assistive devices and communication strategies.

The largest burden of NIHL has been through occupational exposures; however, noise-induced hearing loss can also be due to unsafe recreational, residential, social and military service-related noise exposures. It is estimated that 15% of young people are exposed to sufficient leisure noises (i.e. concerts, sporting events, daily activities, personal listening devices, etc.) to cause NIHL. There is not a limited list of noise sources that can cause hearing loss; rather, exposure to excessively high levels from any sound source over time can cause hearing loss.

Signs and Symptoms

The first symptom of NIHL may be difficulty hearing a conversation against a noisy background. The effect of hearing loss on speech perception has two components. The first component is the loss of audibility, which may be perceived as an overall decrease in volume. Modern hearing aids compensate this loss with amplification. The second component is known as "distortion" or "clarity loss" due to selective frequency loss. Consonants, due to their higher frequency, are typically affected first. For example, the sounds "s" and "t" are often difficult to hear for those with hearing loss, affecting clarity of speech. NIHL can affect either one or both ears. Unilateral hearing loss causes problems with directional hearing, affecting the ability to localize sound.

Temporary and Permanent Hearing Changes

- PTS (Permanent Threshold Shift) is a permanent change of the hearing threshold (the intensity necessary for one to detect a sound) following an event, which will never recover. PTS is measured in decibels.
- TTS (Temporary Threshold Shift) is a temporary change of the hearing threshold the hearing loss that will be recovered after a few hours to couple of days. Also called auditory fatigue. TTS is also measured in decibels.

In addition to hearing loss, other external symptoms of an acoustic trauma can be:

- Tinnitus;
- Otalgia;
- Hyperacusis;
- Dizziness or vertigo; in the case of vestibular damages, in the inner-ear.

Tinnitus

Tinnitus is described as hearing a sound when an external sound is not present. Noise-induced hearing loss can cause high-pitched tinnitus. An estimated 50 million Americans have some degree of tinnitus in one or both ears; 16 million of them have symptoms serious enough for them to see a doctor or hearing specialist. As many as 2 million become so debilitated by the unrelenting ringing, hissing, chirping, clicking, whooshing or screeching, that they cannot carry out normal daily activities.

Tinnitus is the largest single category for disability claims in the military, with hearing loss a close second. The third largest category is post-traumatic stress disorder, which itself may be accompanied by tinnitus and may exacerbate it.

Quality of Life

NIHL has implications on quality of life that extend beyond related symptoms and the ability to hear. The annual disability-adjusted life years (DALYs) were estimated for noise-exposed U.S. workers. DALYs represent the number of healthy years lost due to a disease or other health condition. They were defined by the 2013 Global Burden of Disease (GBD) Study. The DALYs calculation accounts for life limitations experienced because of hearing loss as a lost portion of a healthy year of life. So the results indicate the number of healthy years lost by a group of people over a specific time period.

NIOSH used DALYs to estimate the impact of hearing loss on quality of life in the CDC Morbidity and Mortality Weekly Report article "Hearing Impairment Among Noise-Exposed Workers in the United States, 2003-2012." It reported that 2.5 healthy years were lost each year for every 1,000 noise-exposed U.S. workers because of hearing impairment (hearing loss that impacts day-to-day activities). These lost years were shared among the 13% of workers with hearing impairment (about 130 workers out of each 1,000 workers). Mining, Construction and Manufacturing workers lost more healthy years than workers in other industry sectors; specifically and respectively in those sectors, 3.5, 3.1 and 2.7 healthy years were lost each year for every 1,000 workers.

Negative Impacts

The negative impacts of NIHL on one's ability to reciprocate communication, socialize and interact with society are largely invisible. Hearing loss, in general, is not just an issue of volume; individuals may experience difficulty in understanding what is said over the phone, when several people are talking at once, in a large space, or when the speaker's face cannot be seen. Subsequently, challenging social interactions can negatively lead to decreased self-esteem, shame, and fear. This can be more acutely felt by those who experience hearing impairment or loss earlier in life, rather than later when it is more socially accepted. Such psychosocial states, regardless of age, can lead to social isolation, which is known to negatively impact one's overall health and well-being. The compounding impacts can also lead to depression, especially if hearing impairment leads to tinnitus. Research suggests that those with hearing impairment or loss may be at a greater risk for deterioration of quality of life, as captured by a quote from Helen Keller: "Blindness cuts us off from things, but deafness cuts us off from people." Hearing impairment and loss of hearing, regardless of source or age, also limits experiencing the many benefits of sound on quality of life. In addition to the interpersonal social benefits, new studies suggest the effects of nature sounds, such as birds chirping and water, can positively affect an individual's capacity to recover after being stressed or to increase cognitive focus.

Quality of Life Questionnaire

Hearing loss is typically quantified by results from an audiogram; however, the degree of loss of hearing does not predict the impact on one's quality of life. The impact that NIHL can have on daily life and psychosocial function can be assessed and quantified using a validated questionnaire tool, such as the Hearing Handicap Inventory for the Elderly (HHIE). The HHIE is considered a "useful tool for quantifying the perceived emotional and social/situational consequences of hearing loss." The original tool was designed to test adults 65 years of age and older; however, modified versions exist. For adults the Hearing Handicap Inventory for Adults (HHIA) can be used and for adolescents the modified 28-item Hearing Environments And Reflection on Quality of Life (HEAR-QL-28) can be used. The HHIA, for example, is a 25-item questionnaire that asks both social and emotional-specific questions such as: Does a hearing problem cause you to avoid groups of people?" (social) and "Does a hearing problem cause you to feel frustrated when talking to members of your family?" (emotional). Response options are yes, no and sometimes. A greater score indicates greater perceived handicap.

Cause

The ear can be exposed to short periods of sound in excess of 120 dB without permanent harm — albeit with discomfort and possibly pain; but long term exposure to sound levels over 85 dB(A) can cause permanent hearing loss.

There are two basic types of NIHL:

- NIHL caused by acoustic trauma;
- Gradually developing NIHL.

Acute Acoustic Trauma

NIHL caused by acute acoustic trauma refers to permanent cochlear damage from a one-time exposure to excessive sound pressure. This form of NIHL commonly results from exposure to high-intensity sounds such as explosions, gunfire, a large drum hit loudly, and firecrackers. According to one US study, excessive noise levels in cinemas are sufficiently brief that movie-goers do not experience hearing loss.

Perceived Harmfulness vs. Actual Harmfulness

The discomfort threshold is the loudness level from which a sound starts to be felt as too loud and thus painful by an individual. Industry workers tend to have a higher discomfort threshold (i.e. the sounds must be louder to feel painful than for non-industry workers), but the sound is just as harmful to their ears. Industry workers often suffer from NIHL because the discomfort threshold is not a relevant indicator of the harmfulness of a sound.

Gradually Developing NIHL

Gradually developing NIHL refers to permanent cochlear damage from repeated exposure to loud sounds over a period of time. Unlike acoustic trauma, this form of NIHL does not occur from a single exposure to a high-intensity sound pressure level. Gradually developing NIHL can be caused by multiple exposures to excessive noise in the workplace or any source of repetitive, frequent exposures to sounds of excessive volume, such as home and vehicle stereos, concerts, nightclubs, and personal media players.

Workplace

About 22 million workers are exposed to hazardous noise, with additional millions exposed to solvents and metals that could put them at increased risk for hearing loss. Occupational hearing loss is one of the most common occupational diseases. 49% of male miners have hearing loss by the age of 50. By the age of 60, this number goes up to 70%. Construction workers also suffer an elevated risk. A screening program focused on construction workers employed at US Department of Energy facilities found 58% with significant abnormal hearing loss due to noise exposures at work. Occupational hearing loss is present in up to 33% of workers overall. Occupational exposure to noise causes 16% of adult disabling hearing loss worldwide.

The following is a list of occupations that are most susceptible to hearing loss:

- Agriculture
- Mining
- Construction
- Manufacturing
- Utilities
- Transportation
- Military
- Musicians
- Orchestra conductors

Among Musicians

Musicians, from classical orchestras to rock groups, are exposed to high decibel ranges. Some rock musicians experience noise-induced hearing loss from their music, and some studies have found that "symphonic musicians suffer from hearing impairment and that the impairment might be ascribed to symphonic music."

Music-induced hearing loss is still a controversial topic for hearing researchers. While some population studies have shown that the risk for hearing loss increases as music exposure increases, other studies found little to no correlation between the two. Experts at the 2006 "Noise-Induced Hearing Loss in Children at Work and Play" Conference agreed that further research into this field was still required before making a broad generalization about music-induced hearing loss.

Given the extensive research suggesting that industrial noise exposure can cause sensorineural hearing loss, a link between hearing loss and music exposures of similar level and duration (to industrial noise) seems highly plausible. Determining which individuals or groups are at risk for such exposures may be a difficult task. Despite concerns about the proliferation of personal music players, the there is only scarce evidence supporting their impact on hearing loss, and some small-sample studies suggest that only a fraction of users are affected. People from ages to 6–19 have an approximately 15% rate of hearing loss. Recommendations for musicians to protect their hearing were released in 2015 by NIOSH. The recommendations emphasized education of musicians and those who work in or around the music industry. Annual hearing assessments were also recommended to monitor thresholds as were sound level assessments to help determine the amount of time musicians and related professionals should spend in that environment. Hearing protection was also recommended, and the authors of the NIOSH recommendations even suggested that musicians consider custom earplugs as a way to combat NIHL. Despite these recommendations, musicians continue to face unique challenges in protecting their hearing when compared to individuals in industrial settings. Typically environmental controls are the first line of defense in a hearing conservation program, and depending on the type of musicians some recommendations have been set in place. These recommendations can include adjusting risers or level of the speakers and adjusting the layout of a band or orchestra. These changes in the environment can be beneficial to musicians, but the ability to accomplish them is not always possible. In the cases where these changes cannot be made, hearing protection is recommended. Hearing protection in musicians offers its own sets of benefits and complications. When used properly, hearing protection can limit the exposure of noise in individuals. Musicians have the ability to pick from several different types of hearing protection from conventional ear plugs to custom or high fidelity hearing protection. Despite this, use of hearing protection among musicians is low for several different reasons. Musicians often feel that hearing protective devices can distort how music sounds or makes it too quiet for them to hear important cues which makes them less likely to wear hearing protection even when aware of the risks. Research suggests that educations programs can be beneficial to musicians as well as working with hearing care professionals to help address the specific issues that musicians face.

In 2018, a musician named Chris Goldscheider won a case against Royal Opera House for damaging his hearing in a rehearsal of Wagner's thunderous opera Die Walkure.

Workplace Standards

In the United States, the Occupational Safety and Health Administration (OSHA) describes standards for occupational noise exposure in articles 1910.95 and 1926.52. OSHA states that an employer must implement hearing conservation programs for employees if the noise level of the workplace is equal to or above 85 dB(A) for an averaged eight-hour time period. OSHA also states that "exposure to impulsive or impact noise should not exceed 140 dB peak sound pressure level". The National Institute for Occupational Safety and Health (NIOSH) recommends that all worker exposures to noise should be controlled below a level equivalent to 85 dBA for eight hours to minimize occupational noise induced hearing loss. NIOSH also recommends a 3 dBA exchange rate so that every increase by 3 dBA doubles the amount of the noise and halves the recommended amount of exposure time. The United States Department of Defense (DoD) instruction 605512 has some differences from OSHA 1910.95 standard, for example, OSHA 1910.95 uses a 5 dB exchange rate and DoD instruction 605512 uses a 3 dB exchange rate.

Occupational Noise Exposure Criteria Document.

There are programs that seek to increase compliance and therefore effectiveness of hearing protection rules; the programs include the use of hearing tests and educating people that loud sound is dangerous.

Employees are required to wear hearing protection when it is identified that their eight-hour time weighted average (TWA) is above the exposure action value of 90 dB. If subsequent monitoring shows that 85 dB is not surpassed for an eight-hour TWA, the employee is no longer required to wear hearing protection.

In the European Union, directive 2003/10/EC mandates that employers shall provide hearing protection at noise levels exceeding 80 dB(A), and that hearing protection is mandatory for noise levels exceeding 85 dB(A). Both values are based on 8 hours per day, with a 3 dB exchange rate.

A 2017 Cochrane review found low-quality evidence that legislation to reduce noise in the workplace was successful in reducing exposure both immediately and long-term.

Sporting Events

Several sports stadiums pride themselves in having louder stadiums than their opponents because it may create a more difficult environment for opposing teams to play in. Currently, there are few studies on noise in sports stadiums, but some preliminary measurements show noise levels reaching 120 dB, and informal studies suggest that people may receive up to a 117% noise dose in one game. There are many challenges that face hearing conservationists such as sports culture. Sports fans will create noise in an attempt to distract other teams, and some sports teams have been known to create artificial noise in an attempt to make the stadium louder. In doing that, workers, teams, and fans may be at potential risk for damage to the auditory system.

NIOSH conducted a health hazard evaluation and studies at Monster Trucking and Stock Car racing events, spectators average noise levels ranged from 95 to 100 dBA at the Monster Truck event and over 100 dBA at the stock car racing event. NIOSH researchers also published noise exposure levels for drivers, crew members, and staff. Noise levels at the Bristol Motor Speedway ranged from 96 dBA in the stands to 114 dBA for a driver inside a car during practice. Peak noise levels in the Pit area reached or exceeded 130 dB SPL, a level often associated with human hearing threshold for pain. Several prominent NASCAR drivers have complete or partial hearing loss and other symptoms from their many years of exposure.

During the FIFA World Cup in 2010, noise levels created by fans blowing Vuvuzela averaged 131 dBA at the horn opening and 113 dBA at 2 meter distance. Peak levels reached 144 dB SPL, louder than a jet engine at takeoff.

A study of occupational and recreational noise exposure at indoor hockey arenas found noise levels from 81 dBA to 97 dBA, with peak sound pressure levels ranging from 105 dB SPLto 124 dB SPL. Another study examined the hearing threshold of hockey officials and found mean noise exposures of 93 dBA. Hearing threshold shifts were observed in 86% of the officials.

In a study of noise levels at 10 intercollegiate basketball games showed noise levels in 6 of the 10 basketball games to exceed the national workplace noise exposure standards, with participants showing temporary threshold levels at one of the games.

While there is no agency that currently monitors sports stadium noise exposure, organizations such as NIOSH or OSHA use occupational standards for industrial settings that some experts feel could be applied for those working at sporting events. Workers often will not exceed OSHA standards of 90 dBA, but NIOSH, whose focus is on best practice, has stricter standards which say that when exposed to noise at or exceeding 85 dBA workers need to be put on a hearing conservation program. Workers may also be at risk for overexposure because of impact noises that can cause instant damage. Experts are suggesting that sports complexes create hearing conservation programs for workers and warn fans of the potential damage that may occur with their hearing.

Studies are still being done on fan exposure, but some preliminary findings show that there are often noises that can be at or exceed 120 dB which, unprotected, can cause damage to the ears in seconds.

Mechanisms

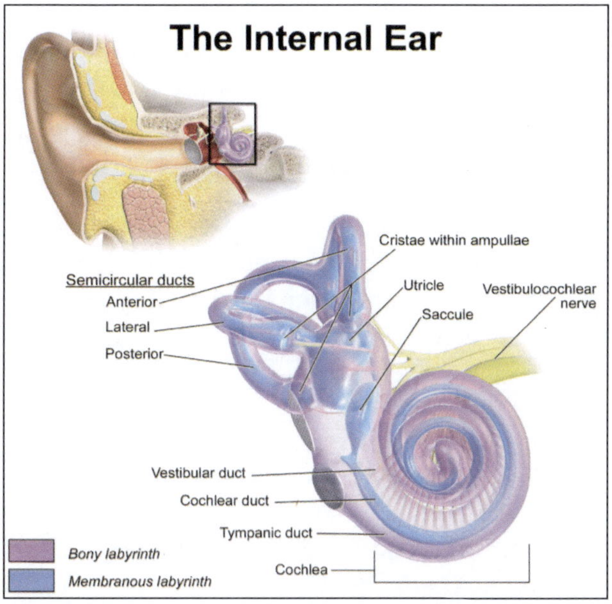

The outer ear receives sound, transmitted through the ossicles of the middle ear to the inner ear, where it is converted to a nervous signal in the cochlear and transmitted along the vestibulocochlear nerve.

NIHL occurs when too much sound intensity is transmitted into and through the auditory system. An acoustic signal from a sound source, such as a radio, enters into the external auditory canal (ear canal), and is funneled through to the tympanic membrane (eardrum), causing it to vibrate. The vibration of the tympanic membrane drives the middle ear ossicles, the malleus, incus, and stapes to vibrate in sync with the eardrum. The middle ear ossicles transfer mechanical energy to the cochlea by way of the stapes footplate hammering against the oval window of the cochlea, effectively amplifying the sound signal. This hammering causes the fluid within the cochlea (perilymph and endolymph) to be displaced. Displacement of the fluid causes movement of the hair cells (sensory cells in the cochlea) and an electrochemical signal to be sent from the auditory nerve (CN VIII) to the central auditory system within the brain. This is where sound is perceived. Different groups of hair cells are responsive to different frequencies. Hair cells at or near the base of the cochlea are most sensitive to higher frequency sounds while those at the apex are most sensitive to lower frequency sounds. There are two known biological mechanisms of NIHL from excessive sound intensity: damage to the structures called stereocilia that sit atop hair cells and respond to sound, and damage to the synapses that the auditory nerve makes with hair cells, also termed "hidden hearing loss".

Physiological Response

The symptoms mentioned are the external signs of the physiological response to cochlear overstimulation. Here are some elements of this response:

- Damaged sensory hairs (stereocilia) of the hair cells; damaged hair cells degenerate and die. In humans and other mammals, dead hair-cells are never replaced; the resulting hearing loss is permanent.

- Inflammation of the exposed areas. This inflammation causes poor blood flow in the exposed blood vessels (vascular stasis), and poor oxygen supply for the liquid inside the cochlea (endolymphatic hypoxia). Those noxious conditions worsen the damaged hair cell degeneration.
- Synaptic damages, by excitotoxicity. Noise over-stimulation causes an excessive release of glutamate, causing the postsynaptic bouton to swell and burst. However the neuron connection can be repaired, and the hearing loss only caused by the "wiring" (i.e. excitotoxicity) can thus be recovered within 2–3 days.

Hair Cell Damage or Death

When the ear is exposed to excessive sound levels or loud sounds over time, the overstimulation of the hair cells leads to heavy production of reactive oxygen species, leading to oxidative cell death. In animal experiments, antioxidant vitamins have been found to reduce hearing loss even when administered the day after noise exposure. They were not able to fully prevent it. Damage ranges from exhaustion of the hair (hearing) cells in the ear to loss of those cells. NIHL is therefore the consequence of overstimulation of the hair cells and supporting structures. Structural damage to hair cells (primarily the outer hair cells) will result in hearing loss that can be characterized by an attenuation and distortion of incoming auditory stimuli.

During hair cell death 'scars' develop, which prevent potassium rich fluid of the endolymph from mixing with the fluid on the basal domain. The potassium rich fluid is toxic to the neuronal endings and can damage hearing of the entire ear. If the endolymph fluid mixes with the fluid on the basal domain the neurons become depolarized, causing complete hearing loss. In addition to complete hearing loss, if the area is not sealed and leakage continues further tissue damage will occur. The 'scars' that form to replace the damaged hair cell are caused by supporting hair cells undergoing apoptosis and sealing the reticular lamina, which prevents fluid leakage. The cell death of two supporting hair cells rapidly expands their apical domain, which compresses the hair cell beneath its apical domain.

Nerve Damage

Recent studies have investigated additional mechanisms of NIHL involving delayed or disabled electrochemical transmission of nerve impulses from the hair cell to and along the auditory nerve. In cases of extreme acute acoustic trauma, a portion of the postsynaptic dendrite (where the hair cell transfers electrochemical signals to the auditory nerve) can rupture from overstimulation, temporarily stopping all transmission of auditory input to the auditory nerve. This is known as excitotoxicity. Usually, this sort of rupture heals within about five days, resulting in functional recovery of that synapse. While healing, an over-expression of glutamate receptors can result in temporary tinnitus, or ringing in the ears. Repeated ruptures at the same synapse may eventually fail to heal, leading to permanent hearing loss.

Prolonged exposure to high intensity noise has also been linked to the disruption of ribbon synapses located in the synaptic cleft between inner hair cells and spiral ganglion nerve fibers, leading to a disorder referred to as cochlear synaptopathy or hidden hearing loss. This disorder is cumulative and over time, leads to degeneration of the spiral ganglion cells of the inner ear and overall dysfunction in the neural transmission between auditory nerve fibers and the central auditory pathway. The most common symptom of cochlear synaptopathy is difficulty understanding speech, especially in the presence of competing noise. However, this type of hearing impairment is often undetectable by conventional pure tone audiometry, thus the name "hidden" hearing loss.

Acoustic over-exposure can also result in decreased myelination at specific points on the auditory nerve. Myelin, an insulating sheath surrounding nerve axons, expedites electrical impulses along nerves throughout the nervous system. Thinning of the myelin sheath on the auditory nerve significantly slows the transmission of electrical signals from hair cell to auditory cortex, reducing comprehension of auditory stimuli by delaying auditory perception, particularly in noisy environments.

Individual Susceptibility towards Noise

There appear to be large differences in individual susceptibility to NIHL. The following factors have been implicated:

- Missing acoustic reflex;
- Previous sensorineural hearing loss;
- A bad general health state: bad cardiovascular function, insufficient intake of oxygen, a high platelet aggregation rate; and most importantly, a high viscosity of the blood;
- Cigarette smoking;
- Exposure to ototoxic chemicals (medication or environmental chemicals that can damage the ear), including certain solvents and heavy metals;
- Type 2 diabetes.

Diagnosis

Both NIHL caused by acoustic trauma and gradually-developed-NIHL can often be characterized by a specific pattern presented in audiological findings. NIHL is generally observed to decrease hearing sensitivity in the higher frequencies, also called an audiometric notch, especially at 4000 Hz, but sometimes at 3000 or 6000 Hz. The symptoms of NIHL are usually presented equally in both ears.

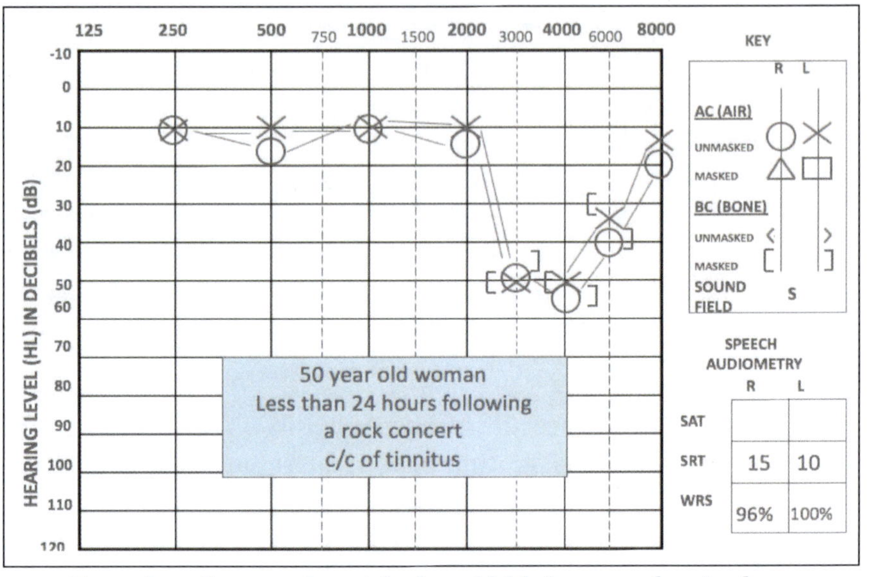

Example audiogram of a notch-shaped high frequency hearing loss.

This typical 4000 Hz notch is due to the transfer function of the ear. As does any object facing a sound, the ear acts as a passive filter, although the inner ear is not an absolute passive filter because the outer hair cells provide active mechanisms. A passive filter is a low pass: the high frequencies are more absorbed by the object because high frequencies impose a higher pace of compression-decompression to the object. The high frequency harmonics of a sound are more harmful to the inner-ear.

However, not all audiological results from people with NIHL match this typical notch. Often a decline in hearing sensitivity will occur at frequencies other than at the typical 3000–6000 Hz range. Variations arise from differences in people's ear canal resonance, the frequency of the harmful acoustic signal, and the length of exposure. As harmful noise exposure continues, the commonly affected frequencies will broaden to lower frequencies and worsen in severity.

Prevention

NIHL can be prevented through the use of simple, widely available, and economical tools. This includes but is not limited to personal noise reduction through the use of ear protection (i.e. earplugs and earmuffs), education, and hearing conservation programs. For the average person, there are three basic things that one can do to prevent NIHL: Turn down the volume on devices, move away from the source of noise, and wear hearing protectors in loud environments.

Non-occupational noise exposure is not regulated or governed in the same manner as occupational noise exposure; therefore prevention efforts rely heavily on education awareness campaigns and public policy. The WHO cites that nearly half of those affected by hearing loss could have been prevented through primary prevention efforts such as: "reducing exposure (both occupational and recreational) to loud sounds by raising awareness about the risks; developing and enforcing relevant legislation; and encouraging individuals to use personal protective devices such as earplugs and noise-cancelling earphones and headphones."

Personal Noise Reduction Devices

Personal noise reduction devices can be passive, active or a combination. Passive ear protection includes earplugs or earmuffs which can block noise up to a specific frequency. Earplugs and earmuffs can provide the wearer with 10 dB to 40 dB of attenuation. However, use of earplugs is only effective if the users have been educated and use them properly; without proper use, protection falls far below manufacturer ratings. Higher consistency of performance has been found with custom-molded earplugs. Because of their ease of use without education, and ease of application or removal, earmuffs have more consistency with both compliance and noise attenuation. Active ear protection (electronic pass-through hearing protection devices or EPHPs) electronically filter out noises of specific frequencies or decibels while allowing the remaining noise to pass through. A personal attenuation rating can be objectively measured by using a hearing protection fit testing system.

Education

Education is key to prevention. Before hearing protective actions will take place, a person must understand they are at risk for NIHL and know their options for prevention. Hearing protection programs have been hindered by people not wearing the protection for various reasons, including the desire to converse, uncomfortable devices, lack of concern about the need for protection, and social pressure against wearing protection. Although youth are at risk for hearing loss, one study found that 96.3% of parents did not believe their adolescents were at risk, and only 69% had talked to their children about hearing protection; those aware of NIHL risks were more likely to talk to their teens.

A systematic review of the effectiveness of interventions to promote the use of hearing protection devices such as earplugs and earmuffs among workers found that tailored interventions improve the average use of such devices when compared with no intervention. Tailored interventions involve the use of communication or other types of interventions that are specific to an individual or a group and aim to change behavior. Mixed interventions such as mailings, distribution of hearing protection devices, noise assessments, and hearing testing are also more effective in improving the use of hearing protection devices compared with hearing testing alone. Programs that increased the proportion of workers wearing hearing protection equipment did reduce overall hearing loss.

Personal Listening Devices

Although research is limited, it suggests that increased exposure to loud noise through personal listening devices is a risk factor for noise induced hearing loss. More than half of people are exposed to sound through music exposure on personal devices greater than recommended levels. Research suggests stronger correlations between extended duration or elevated usage of personal listening devices and hearing loss.

Hearing Conservation Programs

Workers in general industry who are exposed to noise levels above 85 dBA are required by the Occupational Safety and Health Administration (OSHA) to be in a hearing conservation program (HCP), which includes noise measurement, noise control, periodic audiometric testing, hearing protection, worker education, and record keeping. Twenty-four states, Puerto Rico, and the U.S. Virgin Islands have OSHA-approved state plans and have adopted their own standards and enforcement policies. Most of these state standards are identical to those of federal OSHA. However, some states have adopted different standards or may have different enforcement policies. Most health and safety regulations are designed to keep damage risk within "acceptable limits" — that is, some people are likely to incur a hearing loss even when exposed to less than the maximum daily amount of noise specified in a regulation. Hearing conservation programs in other arenas (schools, military) have become more common, and it has been established that unsafe listening behaviors, such as listening to loud noise for extended periods of time without protection, persist despite knowledge of potential hearing loss effects.

However, it is understood that HCPs are designed to change behavior, which is known to be a complex issue that requires a multi-faceted approach. According to Keppler et al. in their 2015 study of such programming, they cite the necessary attitude change towards the susceptibility of risk and degree of severity of hearing loss. Among young adults, the concept of severity is most crucial because it has been found that behavior change may not occur unless an individual experiences NIHL or similarly related NIHL tinnitus, furthering warranting a multi-pronged approach based on hearing conservation programming and education.

Interventions to prevent noise-induced hearing loss often have many components. A 2017 Cochrane review found that hearing loss prevention programs suggest that stricter legislation might reduce noise levels. Giving workers information on their noise exposure levels by itself was not shown to decrease exposure to noise. Ear protection, if used correctly, has the potential to reduce noise to safer levels, but does not necessarily prevent hearing loss. External solutions such as proper maintenance of equipment can lead to noise reduction, but further study of this issue under real-life conditions is needed. Other possible solutions include improved enforcement of existing legislation and better implementation of well-designed prevention programmes, which have not yet been proven conclusively to be effective. The implications is that further research could affect conclusions reached.

Several hearing conservation programs have been developed to educate a variety of audiences about the dangers of NIHL and how to prevent it. Dangerous Decibels aims to significantly reduce the prevalence of noise induced hearing loss and tinnitus through exhibits, education and research. 'We're hEAR for You' is a small non-profit that distributes information and ear plugs at concert and music festival venues. The Buy Quiet program was created to combat occupational noise exposures by promoting the purchase of quieter tools and equipment and encourage manufacturers to design quieter equipment. The National Institute on Deafness and Other Communication Disorders developed the It's a Noisy Planet. Protect their Hearing educational campaign to inform preteens, parents, and educators about the causes and prevention of NIHL. The National Institute for Occupational Safety and Health partnered with the National Hearing Conservation Association in 2007 to establish the Safe-in-Sound Excellence and Innovation in Hearing Loss Prevention Awards to recognize organizations that are successfully implementing hearing loss prevention concepts into their daily routines.

Medication

Medications are still being researched to determine if they can prevent NIHL. No medication has been proven to prevent or repair NIHL in humans.

There is evidence that hearing loss can be minimized by taking high doses of magnesium for a few days, starting as soon as possible after exposure to the loud noise. A magnesium-high diet also seems to be helpful as an NIHL-preventative if taken in advance of exposure to loud noises. Consuming sizable amounts of magnesium can be potentially harmful, so this treatment should be followed with caution.

Sound or Stress Training

Despite different people having different thresholds for what noises are painful, this pain threshold had no correlation with which noises cause hearing damage. The ear can not get more resistant to noise harmfulness by training it to noise. The cochlea is partially protected by the acoustic reflex, but being frequently exposed to noise does not lower the reflex threshold. It had been observed that noise conditioning (i.e. exposure to loud non-traumatizing noise) several hours prior to the exposure to traumatizing sound level, significantly reduced the damages inflicted to the hair-cells. The same "protective effect" was also observed with other stressors such as heat-shock conditioning and stress (by restraint) conditioning. This "protective effect" only happens if the traumatizing noise is presented within an optimum interval of time after the sound-conditioning session (-24 hours for a 15 min. sound-conditioning; no more protection after 48 hours). This "protective effect" had long been thought to involve the active mechanisms of the outer hair cells and the efferent system commanding them.

The contractile effect of the outer hair cells, activated by the efferent nervous system has been proven to provide a protective effect against acoustic trauma.

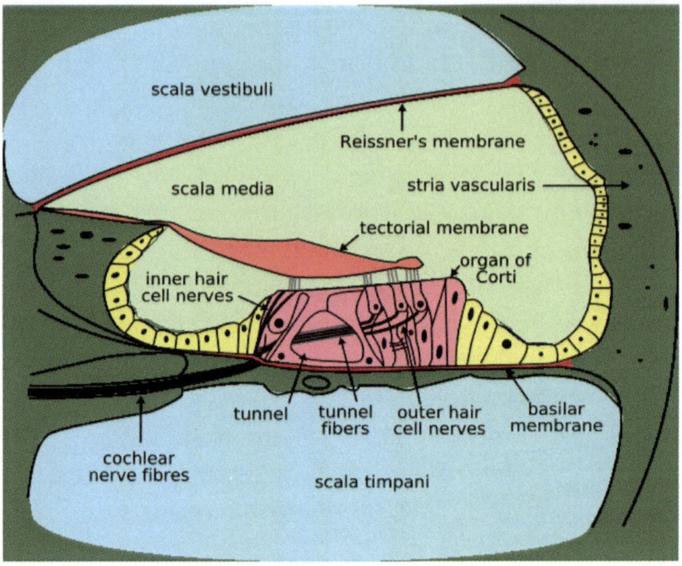

Cross-section of the cochlea. The inner hair cells are connected to afferent nerve fibers, and the outer hair cells are connected to efferent nerve fibers.

However, a 2006 study revealed a different protective mechanism for stress conditioning. The study revealed that the stressor (sound, heat, or stress) conditioning increases the receptibility to glucocorticoid, a kind of anti-inflammatory hormone. The effects of glucocorticoid thus mitigate the inflammation from an acoustic trauma that can lead to hearing loss. In fact, high doses of corticoids are often prescribed by physicians after an acoustic-trauma in order to mitigate the inflammatory response.

Sound (or other stressor) conditioning is a pre-emptive medication against cochlea inflammation. It does not make the ear more resistant to noise. It reduces the inflammation caused by the acoustic trauma, which would cause subsequent damages to hair cells. While an anti-inflammatory medication would increase the quantity of anti-inflammatory hormone in the whole body, noise conditioning increases the number of receptors for the anti-inflammatory hormone, and only in the areas where it is much needed (i.e. cochlea).

Physiological Response

- Stressor (noise, heat shock or stress) conditioning activates hormonal glands: the HPA axis. Note that the HPA axis is associated to the immune system.
- This HPA axis activation results in the up regulation of glucocorticoid receptors (GR) in the cochlea and the paraventricular nucleus (PVN) of the hypothalamus. Note that the glucocorticoid hormone is a kind of immune-reaction-inhibitor, including the inflammation reaction.

- This up regulation of GR thus prevents GR down regulation induced by acoustic trauma.
- The protective effect of noise-conditioning is blocked by adrenalectomy or pharmacological treatment with RU486 + metyrapone (a glucocorticoid receptor antagonist).

Treatment

Treatment options that offer "cures" for NIHL are under research and development. Currently there are no commonly used cures, but rather assistive devices and therapies to try and manage the symptoms of NIHL.

Acute Acoustic Trauma

Several clinical trials have been conducted to treat temporary NIHL occurring after a traumatic noise event, such as a gunshot or firework. In 2007, individuals with acute acoustic trauma after firecracker exposure were injected intratympanically with a cell permeable ligand, AM-111. The trial found AM-111 to have a therapeutic effect on at least two cases of those with acute trauma. Treatment with a combination of prednisolone and piracetam appeared to rescue patients with acute trauma after exposure to gunshots. However, those who received the treatment within an hour of exposure had higher rates of recovery and significantly lower threshold shifts compared to those who received treatment after one hour.

Additionally, clinical trials using antioxidants after a traumatic noise event to reduce reactive oxygen species have displayed promising results. Antibiotic injections with allopurinol, lazaroids, α-D-tocopherol, and mannitol were found to reduce the threshold shift after noise exposure. Another antioxidant, Ebselen, has been shown to have promising results for both TTS and PTS. Ebselen mimics gluthathione peroxide, an enzyme that has many functions, including scavenging hydrogen peroxide and reactive oxygen species. After noise exposure, gluthathione peroxide decreases in the ear. An oral administration of ebselen in both preclinical tests on guinea pigs and human trials indicate that noise induced TTS and PTS was reduced.

Recently, combination therapy with hyperbaric oxygen therapy (HBO) and corticosteroids has been found to be effective for acute acoustic trauma. Acute noise exposure causes inflammation and lower oxygen supply in the inner ear. Corticosteroids hinder the inflammatory reaction and HBO provides an adequate oxygen supply. This therapy has been shown to be effective when initiated within three days after acoustic trauma. Therefore, this condition is considered an ENT emergency.

Gradually Occurring NIHL

At the present time, no established clinical treatments exist to reverse the effects of permanent NIHL. However, current research for the possible use of drug and genetic therapies look hopeful. In addition, management options such as hearing aids and counseling exist.

Many studies have been conducted looking at regeneration of hair cells in the inner ear. While hair cells are generally not replaced through cell regeneration, mechanisms are being studied to induce replacement of these important cells. One study involves the replacement of damaged hair cells with regenerated cells, via the mechanism of gene transfer of atonal gene Math1 to pluripotent stem cells within the inner ear. Other atonal genes are being studied to induce regeneration of hair cells in the inner ear.

Management

For people living with NIHL, there are several management options that can improve the ability to communicate. These options include counseling, amplification, and other assisted listening devices, such as frequency modulation (FM) systems. FM systems can enhance the use of hearing aids and overcome the effects of poor listening conditions because the signal is sent from the microphone worn by the speaker directly to the listener. The prognosis has improved with the recent advancements in digital hearing aid technology, such as directional microphones, open-fit hearing aids, and more advanced algorithms. Hearing aids can mask or cover up the tinnitus, and many with hearing loss and tinnitus find relief by using hearing aids. Though there is no cure or agreed-upon treatment for tinnitus, some drugs have been shown to provide temporary reduction of tinnitus. Other treatments for tinnitus include cognitive-behavioral therapy, biofeedback, and electrical stimulation. Annual audiological evaluations are recommended to monitor any changes in a patient's hearing and to modify hearing-aid prescriptions.

A systematic-review conducted by the American Academy of Audiology Task Force On the Health-Related Quality of Life Benefits of Amplification in Adults found the use of hearing aids to increase quality of life. The review pertained to adults who experienced sensorineural hearing loss, which can be caused by excessive, loud noise.

Epidemiology

The World Health Organization estimates that nearly 360 million people have moderate to profound hearing loss from all causes. Rates of hearing loss has traditionally been attributed to occupational or firearm-related exposure, as well as recreational exposure. The World Health Organization estimated in 2015 that 1.1 billion young people are at risk for hearing loss caused by unsafe listening practices. The over-exposure to excessive loud noise is partially attributed to recreational exposure, such as the use of personal audio devices with music at high volumes for long durations, or social settings such as bars, entertainment and sporting events.

The International Organization for Standardization (ISO) developed the ISO 1999 standards for the estimation of hearing thresholds and noise-induced hearing impairment. They used data from two noise and hearing study databases, one presented by

Burns and Robinson and by Passchier-Vermeer. As race and ethnicity are some of the factors that can affect the expected distribution of pure-tone hearing thresholds several other national or regional datasets exist, from Sweden, Norway, South Korea, the United States and Spain.

In the United States hearing is one of the health outcomes measure by the National Health and Nutrition Examination Survey (NHANES), a survey research program conducted by the National Center for Health Statistics. It examines health and nutritional status of adults and children in the United States. While there is no perfect way to pinpoint hearing loss from excessive noise, researchers look for audiometric notches in a hearing test—dips in the ability to hear certain frequencies—as signs of possible NIHL. As of 2011 data, approximately 24% adults age 20-69 in the United States has an audiometric notch. This data identified differences in NIHL based on age, gender, race/ethnicity, and whether or not a person is exposed to noise at work. Among people aged 20–29, 19.2% had an audiometric notch, compared to 27.3% of people aged 50–59. Males in general had a notch more often than females, regardless of occupational noise exposure, for both unilateral and bilateral audiometric notches. An epidemiological study of 6557 automotive manufacturing workers in China (median age 28 years old) reported that in 62% of the settings where noise exposures were evaluated, levels exceeded the recommended level of 85 dBA. The prevalence of hearing loss was 41% among auto part manufacturing workers, followed by 31% of power train workers and 24% in automotive manufacturing. Across job categories, the highest prevalence rate was observed among welders, of 53%. The prevalance rates were associted with noise levels and the workers' cumulative noise exposure.

Occupational noise exposure is a risk factor for noise induced hearing loss. One study examined hearing test results obtained between 2000-2008 for workers ages 18–65 who had a higher occupational noise exposure than the average worker. Of the sample taken, 18% of the workers had hearing loss. Of the occupations considered, the Mining industry had the highest prevalence and risk of hearing loss, at approximately 27%. Other industries with a higher prevalence and risk included Construction (23.48%) and Manufacturing, especially Wood Product and Non-metallic Mineral Product (19.89%), Apparel (20.18%), and Machinery (21.51%). Estimates of rates of hearing loss have been reported for workers in the Agriculture, Forestry, Fishing, and Hunting (AFFH) sector. The overall prevalence of hearing loss (defined as a pure-tone average threshold across frequencies 1000, 2000, 3000, and 4000 Hz of 25 dB or more in either ear) was 15% but that rate was exceeded in several of the subsectors of those industries. Prevalences were highest among workers in Forest Nurseries and Gathering of Forest Products at 36% and Timber Tract Operations at 22%. The Aquaculture subsector had the highest adjusted risk (adjusted probability ratio of 1.7) of all subsectors of the Agriculture, Forestry, Fishing, and Hunting industries. The same methodology was used to estimate the prevalence of hearing loss for noise-exposed U.S. workers within the Health Care and Social Assistance sector. The prevalence of hearing loss in the Medical Laboratories subsector was 31% and in the Offices of All Other Miscellaneous Health Practitioners subsector was 24%. The Child Day Care Services subsector had a 52% higher risk than the reference industry. While the overall HSA sector prevalence for hearing loss was 19%, the prevalence in the Medical Laboratories subsector and the Offices of All Other Miscellaneous Health Practitioners subsector were 31% and 24%, respectively. The Child Day Care Services subsector had a 52% higher risk than the reference industry of workers who are not exposed to noise at work (Couriers and Messengers).

ACOUSTIC TRAUMA

Acoustic trauma is an injury to the inner ear that's often caused by exposure to a high-decibel noise. This injury can occur after exposure to a single, very loud noise or from exposure to noises at significant decibels over a longer period of time.

Some injuries to the head can cause acoustic trauma if the eardrum is ruptured or if other injuries to the inner ear occur.

The eardrum protects the middle ear and inner ear. It also transmits signals to the brain by way of small vibrations.

Acoustic trauma can damage the way that these vibrations are handled, resulting in hearing loss. Sound moving into the inner ear can cause what doctors sometimes call a threshold shift, which can trigger hearing loss.

Types of Acoustic Trauma

If your doctor believes that your symptoms indicate acoustic trauma, they may try to differentiate between trauma that occurred suddenly through injury and trauma that occurred through ongoing exposure to loud noises.

Different degrees of acoustic trauma can require different treatments.

People at an increased risk for acoustic trauma include those who:

- Work at a job where loud industrial equipment operates for long periods of time.
- Live or work where other high-decibel sounds continue for long periods of time.
- Frequently attend music concerts and other events with high-decibel music.
- Use gun ranges.
- Encounter extremely loud sounds without proper equipment, such as earplugs.

People continually exposed to noise levels over 85 decibels are at an increased risk for acoustic trauma.

Your doctor may provide an estimate of the decibel range of normal daily sounds, like an estimate of around 90 decibels for a small engine. They'll do this to help you assess whether the sounds that you encounter put you at a higher risk for acoustic trauma and hearing loss.

Under 70 decibels or less is considered safe for ongoing listening. This is the estimated noise level of an average group conversation.

Three important factors have a role in acoustic trauma. These include:

- The intensity of sound measured in decibels.
- The pitch or frequency of the sound (higher frequencies are more damaging).
- The total time the person was exposed to the sound.

Symptoms of Acoustic Trauma

The main symptom of acoustic trauma is hearing loss. Injury occurs at the level of the inner ear. The sensitive hair cells can lose their connections to the nerve cells responsible for hearing.

Ear structures may also be directly damaged by loud noise. Sudden sounds above 130 decibelsTrusted Source can damage the ear's natural microphone, the organ of Corti. Acoustic injury can injure the eardrum, along with the small muscles in the ear, particularly the tensor tympani muscle.

In many cases of long-term sound damage, people first begin to have difficulty hearing high-frequency sounds. Difficulty hearing sounds at lower frequencies may occur later. Your doctor may test your response to different frequencies of sound to assess the extent of acoustic trauma.

One of the most important symptoms that can signal the onset of acoustic trauma is called tinnitus. Tinnitus is a type of injury to the ear that causes a buzzing or ringing sound. Those with mild to moderate tinnitus will most often be aware of this symptom when they're in silent environments.

Tinnitus can be caused by drug use, changes to blood vessels, or other conditions and factors, but it's often a precursor to acoustic trauma when it's caused by exposure to loud noises. Tinnitus can be persistent or chronic. Long-term tinnitus is a good reason to suspect acoustic trauma.

Diagnosing Acoustic Trauma

Your doctor will ask you what kind of noises you've been exposed to during different times of your life to help make a diagnosis.

They may also use something called audiometry to detect signs of acoustic trauma. In this test, you're exposed to sounds of varying loudness and of different tones to more carefully assess what you can and can't hear.

Treating Acoustic Trauma

Technological Hearing Assistance

Hearing loss can be treated, but it can't be cured.

Your doctor may recommend technological assistance for your hearing loss condition, such as a hearing aid. New types of hearing aids called cochlear implants may also be available to help you deal with hearing loss from acoustic trauma.

Ear Protection

Your doctor will most likely recommend using earplugs and other kinds of devices to protect your hearing.

These items are part of the personal protective equipment (PPE) that employers should offer to people when they're in a workplace with exposure to loud noises.

Medications

Your doctor may prescribe oral steroid medications to help with some cases of acute acoustic trauma.

However, if you're experiencing hearing loss, your doctor will stress noise protection of the ear and limit exposure to loud environments to prevent the problem from getting worse.

References

- Causes-and-effects-of-noise-pollution: conserve-energy-future.com, Retrieved 19 April, 2019
- Wang y, liberman mc (march 2002). "restraint stress and protection from acoustic injury in mice". Hearing research. 165 (1–2): 96–102. Doi:10.1016/s0378-5955(02)00289-7. Pmid 12031519
- Dealing-with-the-effects-of-noise, noise-pollution, science: britannica.com, Retrieved 5 February, 2019
- Ganguli t (15 february 2009). "hearing loss an inevitable part of racecar driving". Orlando sentinel. Retrieved 9 july 2018
- Acoustic-trauma takeaway, health: healthline.com, Retrieved 26 July, 2019
- Adoga aa, obindo tj (2013). The association between tinnitus and mental illness, mental disorders - theoretical and empirical perspectives. Intech. Isbn 978-953-51-0919-8

5

Noise Pollution: Reduction and Control

Noise pollution can be controlled using various techniques and systems such as soundproofing, radio noise reduction system, acoustiblok, noise barrier, muzzle shroud, hearing protection device, hush kit, acoustic quieting, etc. The chapter closely examines these methods of controlling and reducing noise.

NOISE POLLUTION CAUSES AND CONTROL TECHNIQUES

Here number of noise control principles are presented in the form of general techniques with a practical example for each. These deal both with the factors which influence the generation of noise and its propagation in materials, structures and rooms. First and foremost, noise must be limited at source. When solving a noise problem, the question, "What is the cause of noise disturbance?" has to be answered. The technique described most often uses that method of noise reduction which directly reduces noise generation. Most techniques have a limited area of use, though, and may not be available for all possible situations.

Noise can be caused by a large number of factors, and extensive measurements may often be required before a decision can be made as to which measure should be tried first. It is not usually sufficient to employ just one of the techniques shown to reduce noise. In most cases several different actions need to be taken to achieve the desired result.

In order to make the drawings easy to understand, different symbols are used, e. g. large arrows to indicate high noise radiation and small ones to indicate less noise, wavy lines to indicate structure-borne noise, etc. In a particular case only those symbols which clarify the particular problem under consideration are used.

Rapid Changes Produce Higher Dominant Frequencies

The dominant frequency of the noise produced by an impact is dependent upon the speed of the force, pressure, or velocity change which gives rise to the noise. A rapid change produces a shorter pulse which has higher dominant frequencies. The speed of this change is often determined by the resilience of the two impacting surfaces: The more they deform, the longer they are in contact and the lower the dominant frequencies are. When bouncing a basketball on the floor, the ball is in contact with the floor for a relatively long time. The dominant frequency is therefore low. When playing table tennis the ball is in contact with the bat or table for only a very short time. The dominant frequencies are therefore much higher.

Principle:

Example: A sonic boom lasts approximately as long as it takes for the aircraft to fly through its own length. This may take e.g. 0.25 s. The dominant frequency is therefore about 4 Hz. A gunshot may last only 1 ms. Its dominant frequencies are therefore much higher, at about, kHz.

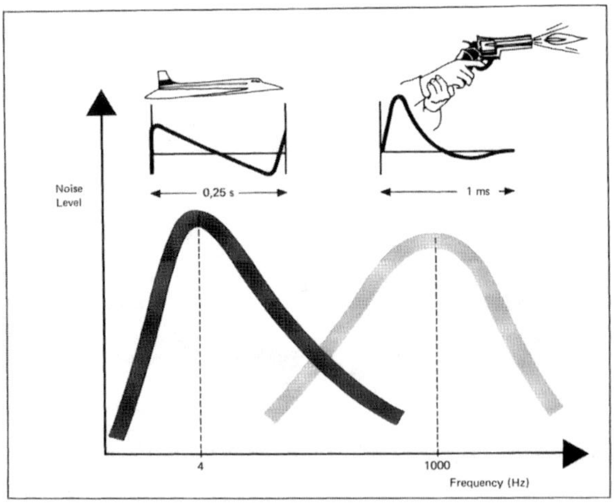

Slow Repetitions give Low Frequencies and Fast Repetitions give High Frequencies

A noise-producing event which repeats, generates frequencies which depend on the time between repetitions. A slowly repeating event gives rise to predominantly low frequencies and a rapidly repeating event gives rise to high frequencies. The level of the sound depends on the magnitude of the change which gave rise to it.

Principle:

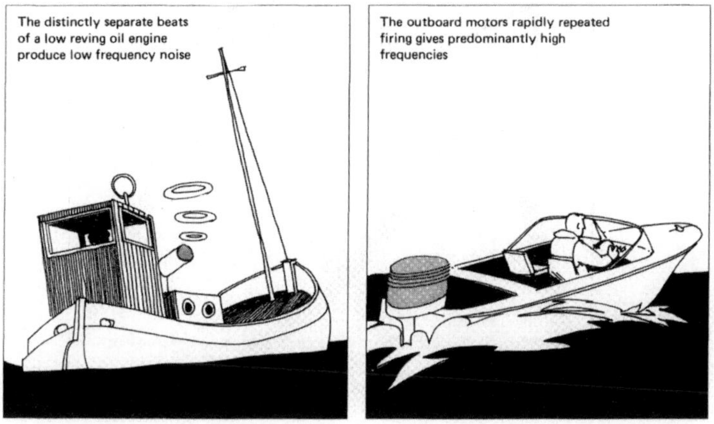

Example: Two gearwheels having the same size but one having twice as many teeth as the other will have predominant frequencies a factor of two apart. The main source of noise is the contact of one tooth on the corresponding tooth on the gearwheel in mesh with it. For the same diameter and speed of rotation the gearwheel with twice as many teeth will have twice as many tooth contacts per second and therefore radiate noise at twice the frequency of the other.

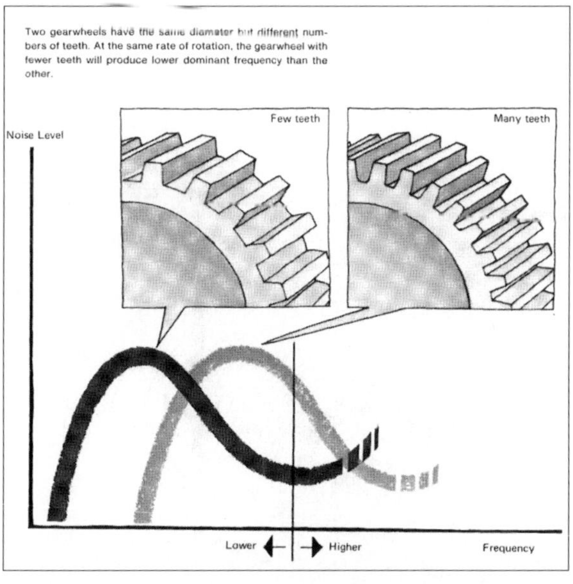

Low Frequency Sound Bends Round Obstacles and through Openings

Low frequency sound radiates approximately equally in all directions. It diffuses round edges and through holes without losing intensity, and reradiates from the edge or from the hole as if it were a new source, again equally in all directions. For this reason screens and barriers are not very effective against it unless they are very large.

Principle:

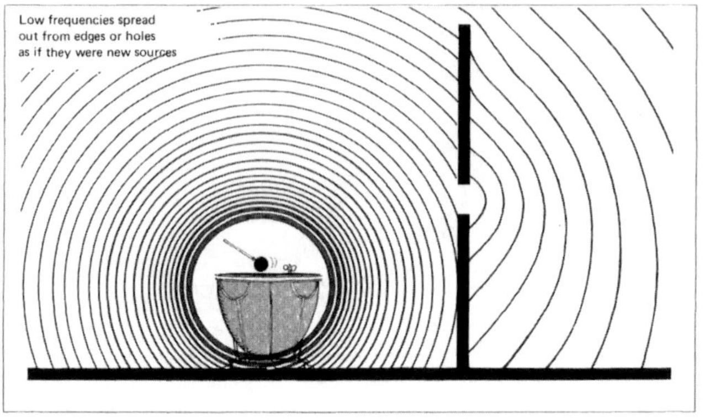

Example: Diesel driven compressors produce high levels of low frequency noise, even if they are furnished with efficient intake and exhaust silencers. Partly open or louvred covers for the intake of cooling air are of little use as noise attenuators. Noise easily radiates out through the openings and gaps.

Solution: Effective quietening of a powerful compressor requires a well-sealed cover eliminating air and noise leaks. The cover can be constructed as a double wall containing ducts with sound absorbent linings. Air for the compressor, for the engine, and for cooling purposes is carried through these ducts, entering and leaving via acoustic louvres. The exhaust silencer is also enclosed within the outer cover. All inspection hatches and access panels must also be tightly fitting and well-sealed.

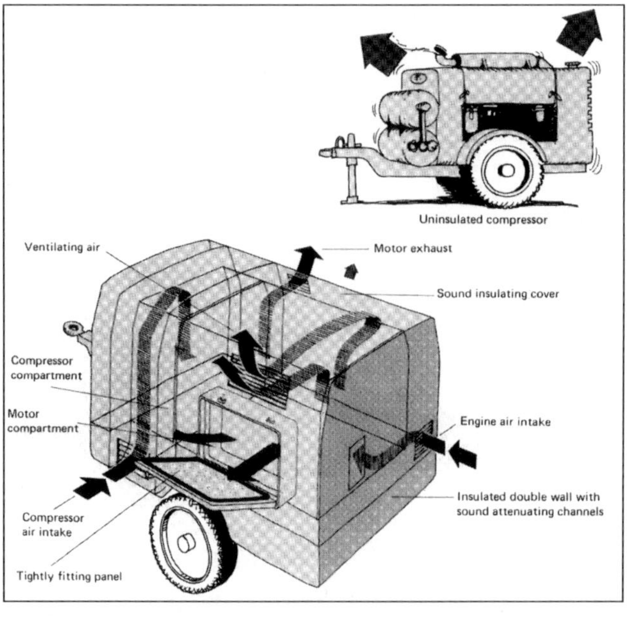

High Frequency Sound is Highly Directional and Easy to Reflect

High frequency sound is often produced by sources which radiate a high noise level in some directions but low levels in others. It can be reflected from a hard surface just as light is reflected by a mirror, and passes through holes in a panel like a beam without being diffused to the sides. Also it cannot diffuse around edges, so barriers are effective against it.

Principle:

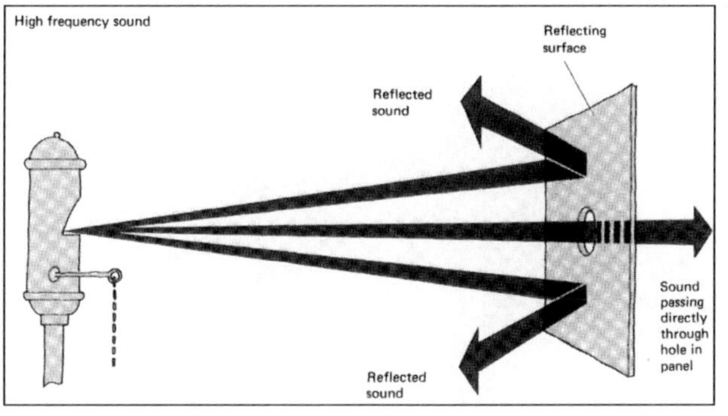

Example: Noise from processes which involve punching, hammering, or other forms of impact give rise to high levels of high frequency noise which can be dangerous to the operator.

Solution: A local enclosure built around the noise source with an opening for access and safety glass to view the work protects the operator's ears from direct sound from the machine. Reflections from the safety glass and most direct sound in other directions are absorbed by the absorbent lining. The small area open for access emits sound only away from the operator's ears.

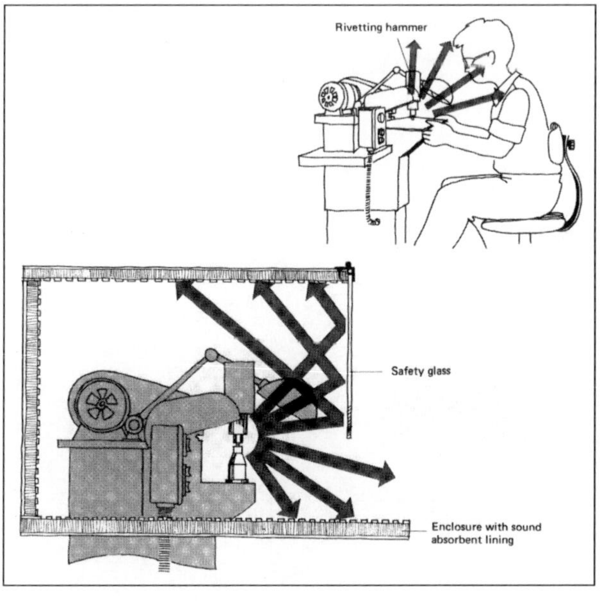

Close to the Source High Frequency Noise Annoys more than Low Frequency Noise

The ear is more sensitive to high frequencies than to low frequencies; so to produce the same amount of annoyance, a low frequency noise must have a higher sound level than a high frequency noise. In some circumstances it may be possible to reduce the annoyance of a close noise source by moving the dominant sound energy to lower frequencies.

Principle:

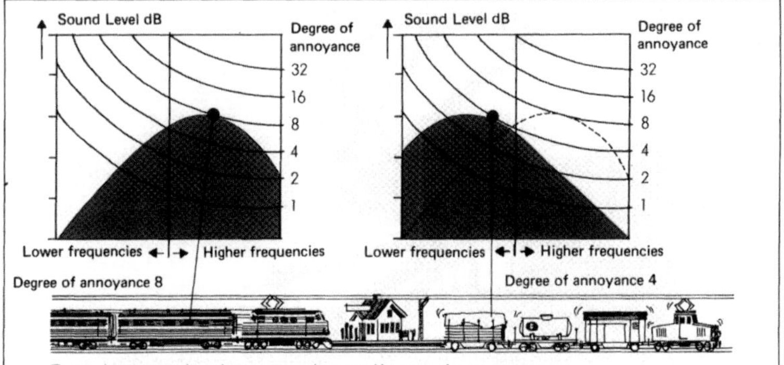

Two trains may produce the same sound energy. However, the passenger train's dominant frequencies are higher than those from the goods train and therefore more annoying.

Example: A ships propeller turns at the same speed as the motor, 125 revolutions per minute, and is the source of most noise on board.

Solution: A design with a larger propeller driven through a gearbox at a lower speed lowers the dominant frequencies and reduces the annoyance caused.

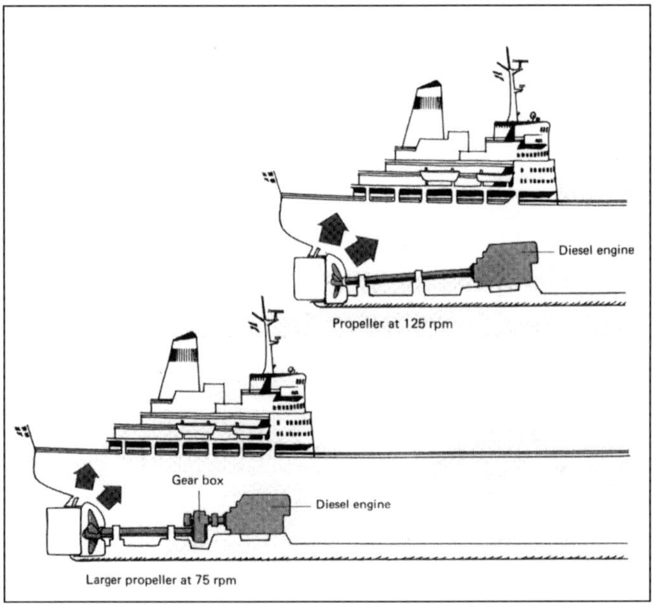

Far from the Source High Frequency Noise Annoys Less than Low Frequency Noise

High frequency noise is attenuated much more, by absorption in air, than low frequency noise is over large distances. This is because absorption is dependent on the number of cycles, and there are more cycles of a high frequency sound than a low frequency in a given distance. In addition, it is normally easier to reduce or shield a source of high frequency noise. If noise in the vicinity of the source is not a problem, it may be possible to shift the dominant noise to a higher frequency which is effectively absorbed by the time it reaches the problem area.

Principle:

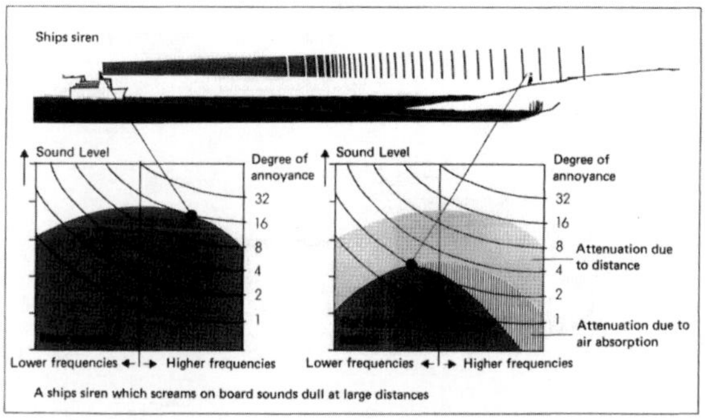

A ships siren which screams on board sounds dull at large distances

Example: Low frequency noise from industrial fans causes noise annoyance in a distant residential area. High frequencies are attenuated on the way.

Solution: The fans can be exchanged with a type with more blades which shift the major sources of noise up in frequency. The higher tones are absorbed sufficiently by the atmosphere so that they are not a source of annoyance in the residen-tiai area, and the low frequency tones are no longer produced. The noise is also easier to attenuate at source.

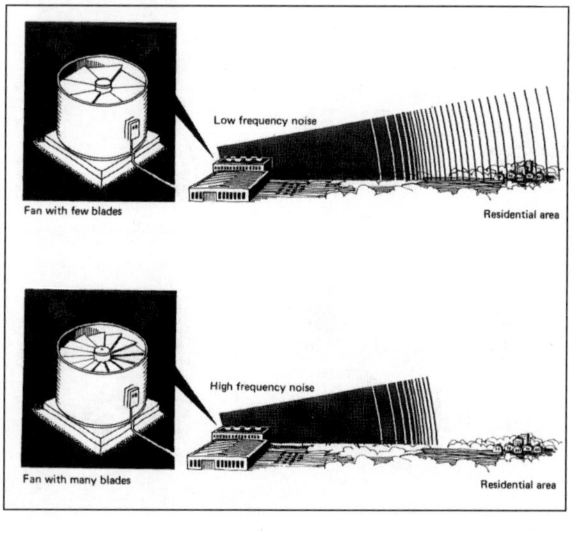

Sound Sources should be Sited away from Reflecting Surfaces

The closer a sound source is placed to a reflecting surface, the more of the sound radiated is directed back into the room. The worst position is against three surfaces, i. e. in a corner. The best position is free-hanging: away from all reflecting surfaces.

Principle:

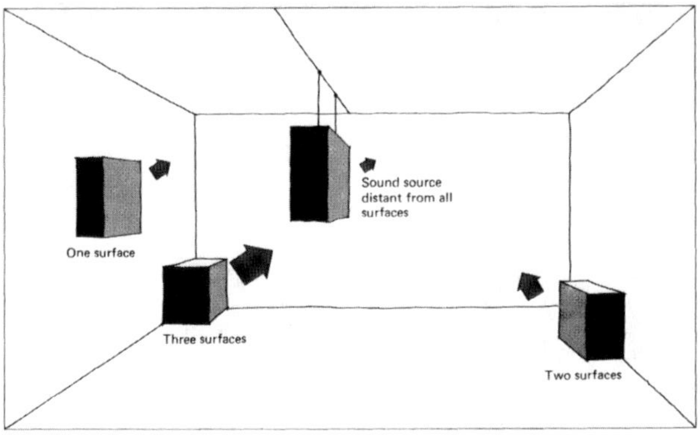

Example: In a machine hall a number of machine tools are placed in four lines, two of them against the walls, with three access lanes between them. This increases the noise from the two lines of machines placed next to the walls.

Solution: The machines along the walls are moved beside the other two lines so there are only two lines. The space along the walls is used as access lanes, of which there are still three, and the overall noise in the hall is reduced.

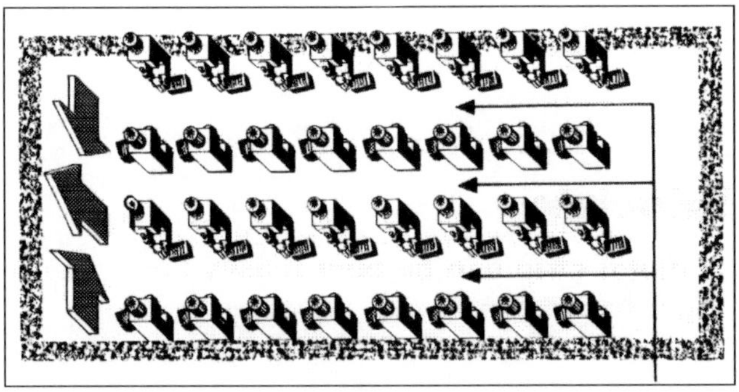

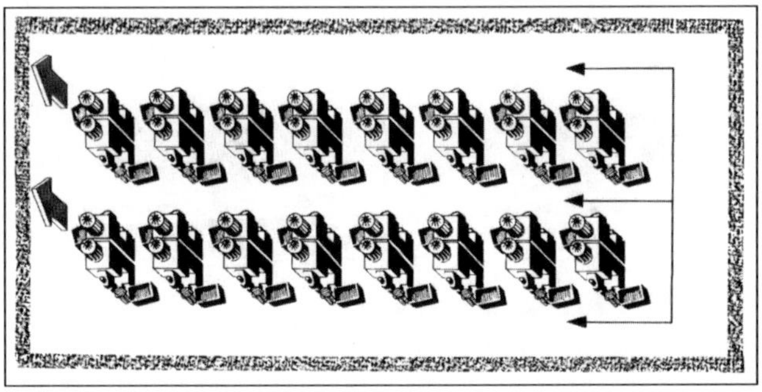

Changes in Force, Pressure or Speed which Lead to Noise

Noise always occurs where there is a change of force, pressure, or speed. Large changes produce the greatest noise, small changes produce less. In many cases the same result can be achieved either with the application of high power over a short period or with less power for a longer period. The first case causes high noise levels, the second, where the power required is small, produces much lower noise levels.

Principle:

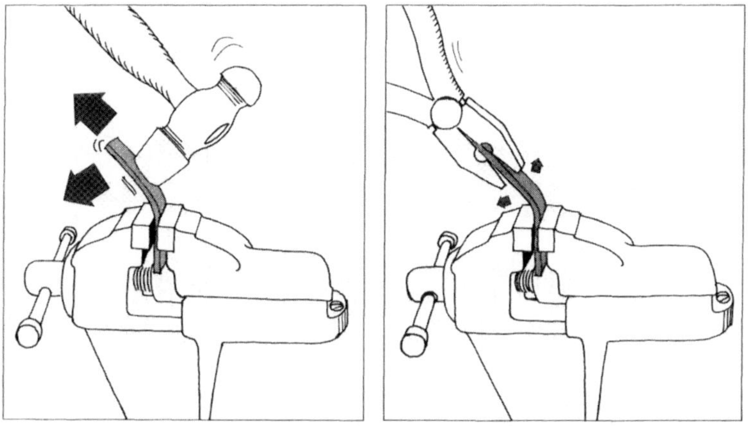

A metal strip can be bent noisily using a hammer or quietly using pliers.

Example: Panels and sheets can be fastened in a number of different ways, some of them very much noisier than others. Those involving impact e.g. nails and rivets are particularly bad from the point of view of hearing damage as they produce very high peak levels of noise.

Solution: In many cases quiet methods such as screws and bolts can be substituted directly without loss of effectiveness or increase in cost, and with the advantage of improved access and ease of dismantling at a later date.

Two Panels can be Fixed Together

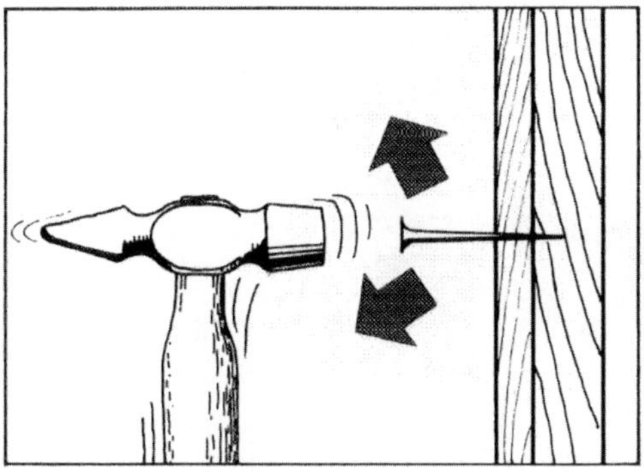

Using nails – noisy.

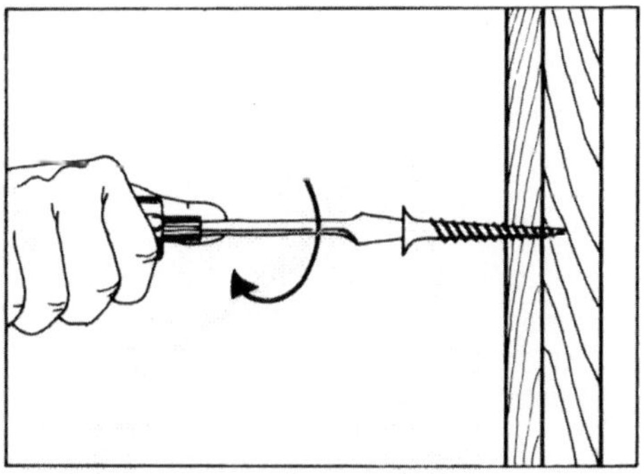

Or screws - quiet.

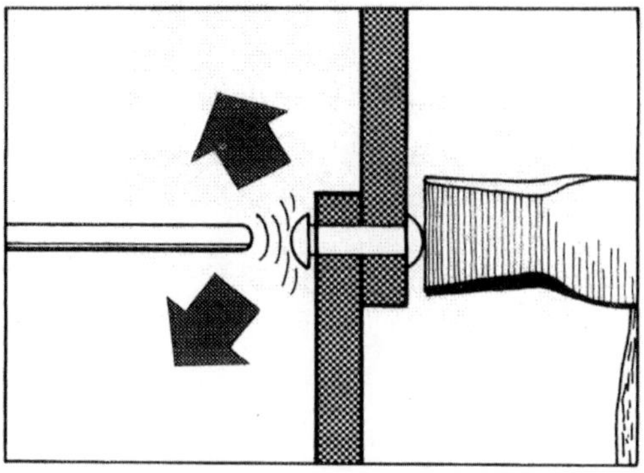

Steel sheet may be rivetted Very noisy.

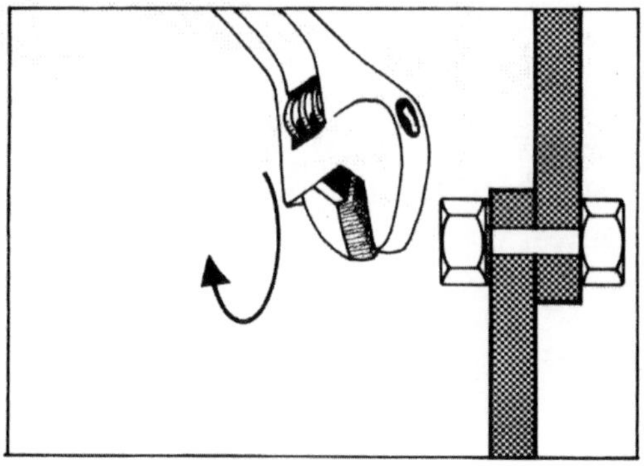

Or bolted - very quiet.

Example: Paving breakers have traditionally been handheld and usually pneumatically powered to reduce weight. High levels of impulse noise are produced both by the chipping process itself and from the exhaust. The operator is exposed to high levels of both noise and vibration.

Solution: A tractor-mounted hydraulic ram driving a hammer can exert a very large static force as well as vibrate. The paving is fractured and the cracked surface can then be levered up by a bucket loader. The noise levels are lower and the operators are further from the source, often in noise protecting cabins.

Example: Cardboard in a carton machine is chopped using a guillotine The knife must fall very quickly using high power in order to cut perpendicular to the production line, causing high noise levels.

Solution: Using a knife which is driven across the production line, the material can be cut with a low force over a longer period, virtually silently. The knife must be set at an angle to the moving line of board to cut perpendicular to the direction of motion.

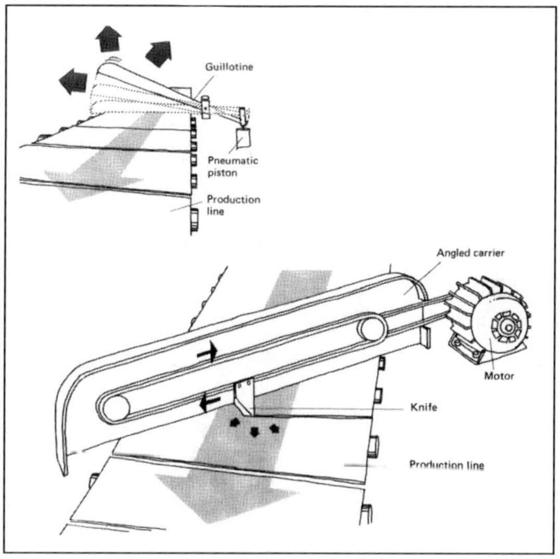

Low Mass and Low Fall Heights give Least Sound

The noise level generated when a panel is struck by a falling object depends primarily on the mass and velocity of the object. The greater the mass and fall height, the louder the noise, because greater energy is available for transfer into the panel via the impact. A reduction in height or in mass by a factor of ten reduces the noise generated by approximately 10 dB.

Principle:

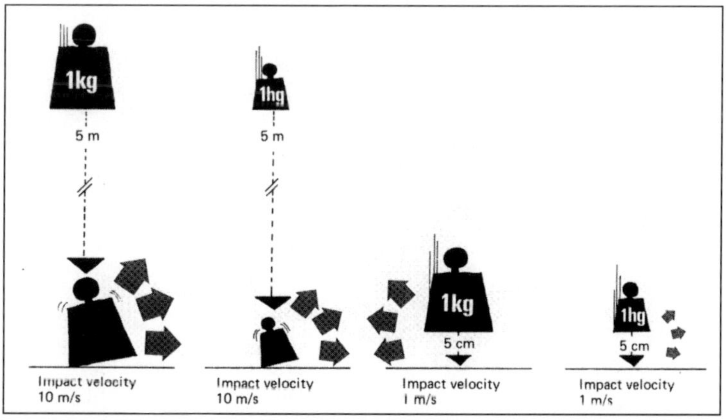

Example: Manufactured items are carried from the producing machine by conveyer and dropped into a collecting bin from a fixed height. When the bin is empty the fall height is large and the noise level is therefore high. There may also be danger of damaging the items.

Solution: The conveyer is constructed so that its height can be adjusted, and is supplied with a case with a number of rubber flaps inside to break the fall of the material. The fall height is therefore never greater than the distance from the collected material to the lowest rubber flap, the conveyer rising automatically as the bin fills.

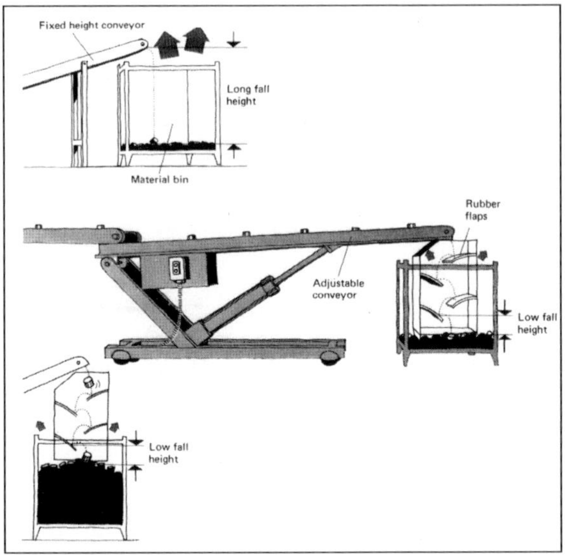

Problem: A material conveyer feeding a hopper deposits the material in the centre of the hopper and the fa" height is therefore large. The hopper itself is also a very resonant structure.

Solution: Mount the conveyer so that the material falls on the edge of the hopper so that the free height is minimised. The interior of the hopper can be lined with wear-resistant material to absorb the impact better, and the external surfaces can be mounted with damping sheets to reduce resonances even further.

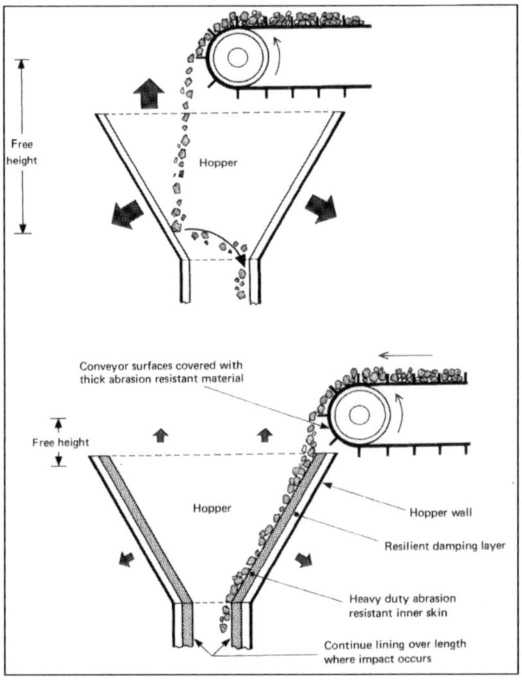

Problem: Sheet piles are normally driven via the impact of a heavy mass dropped from a great height, often powered up again by exploding a diesel charge. Dangerous local noise levels are generated both by the impact on the pile and from the explosion in this case, and annoyance may be caused at distances of up to several kilometers.

Solution: In many situations it is possible to use a completely different technique which avoids impact completely. A set of hydraulically operated rams grip a number of sheet piles simultaneously.

One pile is forced down at a time while the machine pulls upwards on all the rest, which anchor it to the ground. Vibration of the ram holding the pile being driven assists its progress. Impact is avoided completely and noise levels are as low as the hydraulic equipment allows.

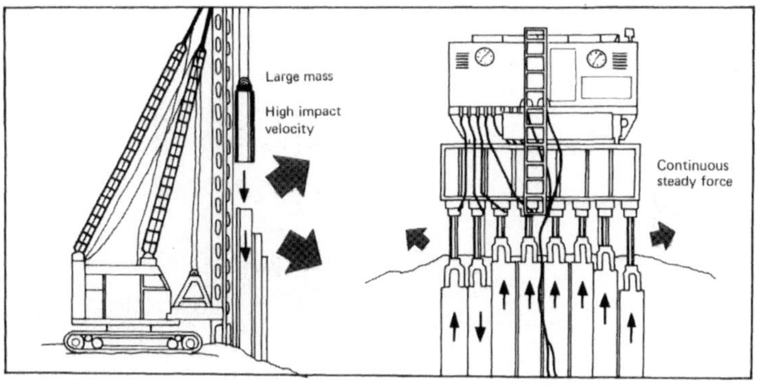

Structure-borne Sound Travels Long Distances

Vibration which gets into a structure, especially homogenous structures such as concrete buildings or ships, travels a very long way because of the very low internal damping of the structure. The energy does not reduce and as soon as a large surface, which acts as a loudspeaker, is connected to the vibrating structure, a high noise level is generated. It is best to isolate the structure from the source of vibration as near to the source as possible.

Principle:

Example: Vibration and stop/start shocks from an elevator can be heard throughout a building. The sound is carried for large distances virtually unattenuated via the concrete slabs.

Solution: The winding machinery must be isolated completely from the building structure using a spring support. Further reduction can be achieved by building the lift shaft and driving mechanism separately from the rest of the building structure.

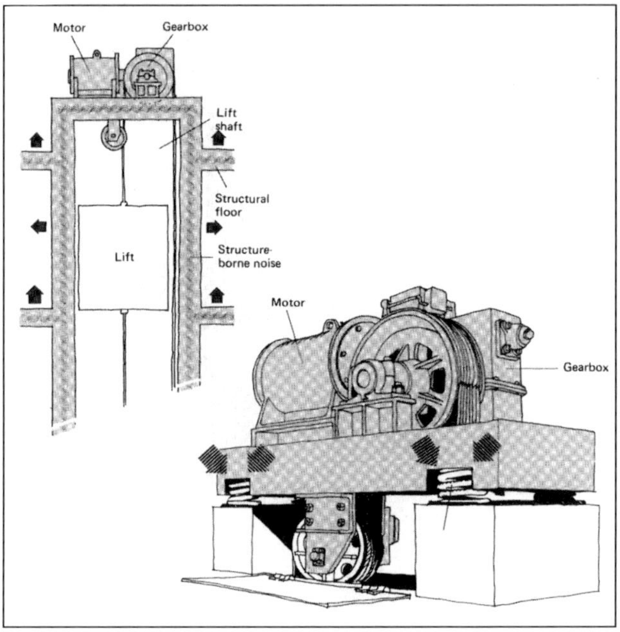

Structure-borne Vibration needs Large Areas to Convert it to Airborne Sound

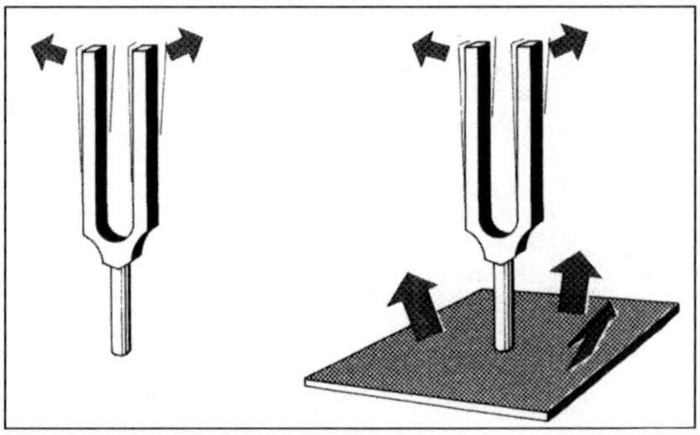

The vibration of a small object will not generally give a high noise level because the area of air set in motion by the object will also be small. The vibration is thus badly matched to the air. However, connecting a large panel transfers the vibration energy into airborne sound much more effeciently by spreading the vibration over a much greater area which gives a high noise level. A tuning fork generates hardly any noise unless connected to a "sounding board". The circulation pump of a central heating system causes the pipework to vibrate, but little noise is transmitted until a large panel in the form of a radiator is connected; radiating not only heat, but noise as well.

Principle:

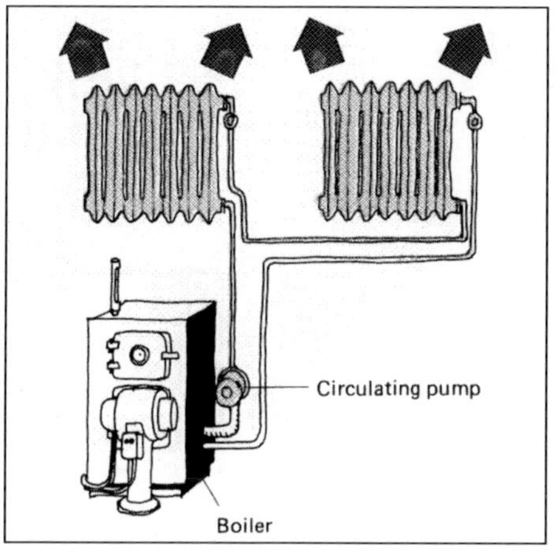

Example: Structure-borne sound in a pipe, perhaps vibration from the circulation pump or noise from the fluid itself, has little opportunity to develop airborne sound as it is of small area. Fixing the pipe to a wall or panel gives the vibration a chance to excite a large area and therefore generate a high airborne sound level.

Solution: The pipework must be properly mounted and isolated from the wall or panels so that they are not set into vibration. This may be done using one of a number of different types of isolator employing springs, rubber strips, foam rubber washers, etc.

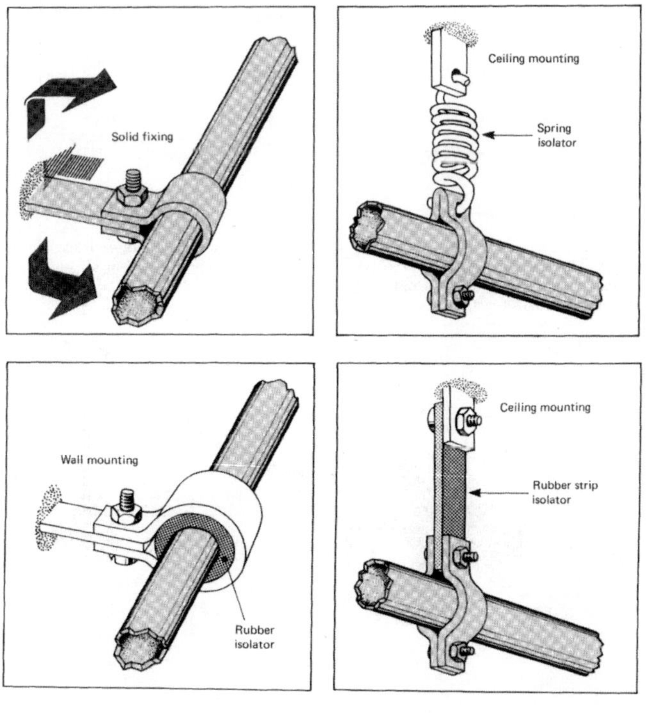

Small Vibrating Objects Radiate Less Noise than Large

A small object may vibrate without giving rise to high noise levels because the surface cannot transfer the vibration energy into sound energy efficiently. Connecting a large panel to the object increases its ability to convert vibration to sound. As most machines produce some vibration, the size of the machine and its panels should be kept as small as possible.

Principle:

The shaver's vibration is transmitted to the glass shelf which vibrates over a large area, amplifying the noise substantially. The vibration is no longer transmitted and the noise is reduced.

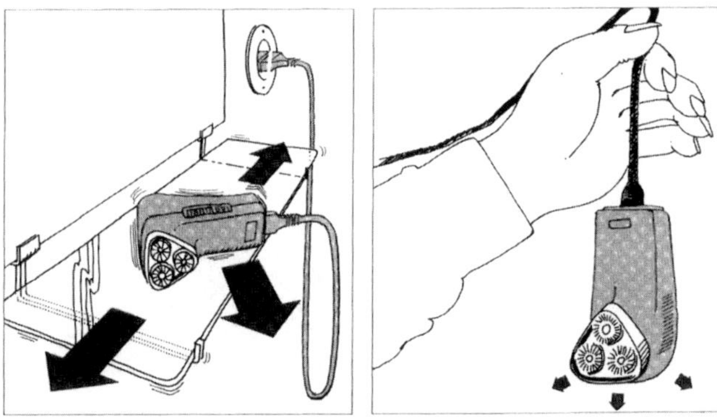

Example: A hydraulic supply system was a significant sound source even though the panels of the oil tank were damped by the oil inside. The chief source of noise was found to be the instrument panel which was set into vibration by the motor.

Solution: Removal of the panel from the machine uncoupled the source of sound from the source of vibration and reduced the sound level.

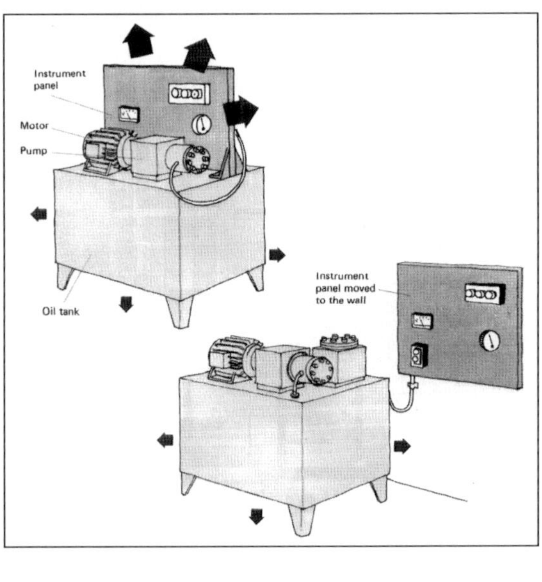

Vibrating Machinery or Parts of Machinery should be Mounted on a Heavy Foundation

Tapping on a light partition generates noise because the partition is easily moved by the force of the tap and therefore transmits the sound. Tapping on a heavy masonary wall produces little noise because the force available is so small it cannot have much effect on the wall. To avoid noise transmission from motors, pumps, etc., they should not be mounted on the relatively flexible equipment which they serve but separately on heavy bases where possible.

Principle:

Example: Pumps and motors serving large pieces of equipment such as hydraulic presses, machine toofs, and turbines are often mounted directly on structural panels. These are set into vibration, radiating high noise levels from the entire area of the machine.

Solution: The services should be mounted, on isolators, away from the main frame of the equipment, on a solid floor where ever possible. Pipework carrying fluids should be connected via flexible piping and include attenuators to avoid the transfer of vibration via these connections back to the main structure of the equipment.

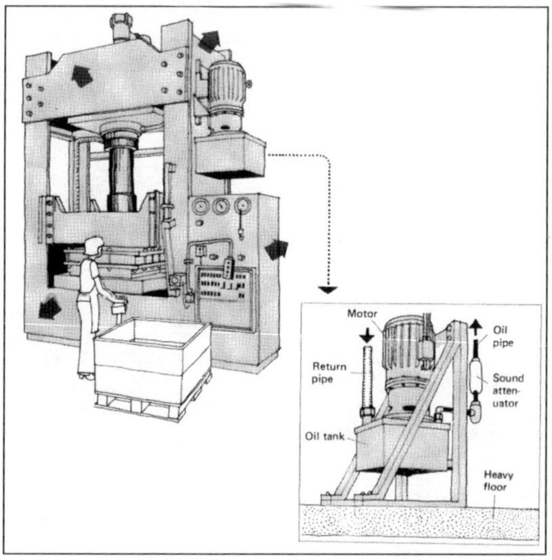

Free Edges on Panels Allow Pressure Equalization around them and Reduce Radiated Noise Levels

It is not always possible to avoid the use of large vibrating panels which give rise to high noise levels. In many cases these may be replaced by a perforated panel or another type with a broken surface. A plain panel radiates noise from all its area efficiently as there are only four sides along which the sound pressure can be partially cancelled out by the negative pressure from the other face. If the panel is perforated, not only is there less surface to radiate the sound, but there are far greater possibilities for this equalization to take place. Noise levels are therefore reduced substantially. Mesh, or expanded metal panels can also be used. For the same reason, a narrow panel radiates less noise than a square panel of the same area.

Principle:

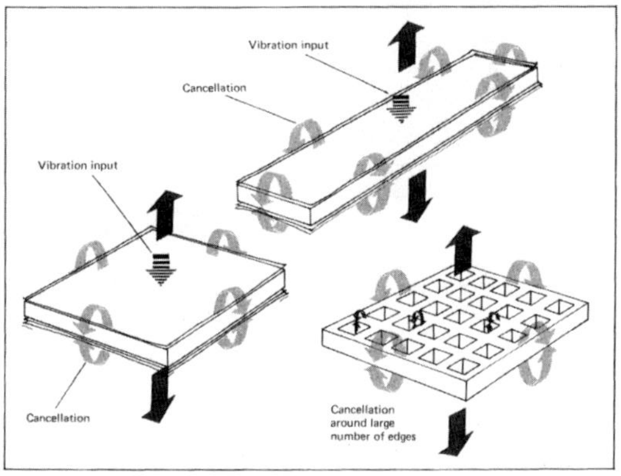

Example: The protective cover over the flywheel and belt drive of a punch press radiates noise efficiently.

Solution: A replacement cover of wire mesh reduces the noise radiation.

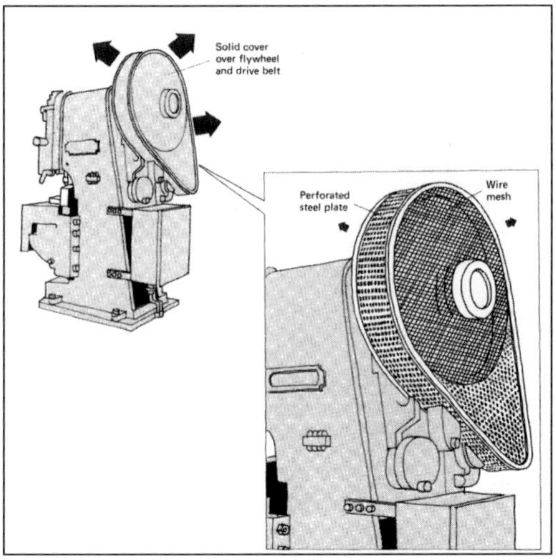

Example: Vibration of wide drive belts on industrial drives can lead to high levels of low frequency noise.

Solution: Replacing the single drive belt with a number of narrower drive belts with gaps between them increases the amount of cancellation which is possible between the top and bottom of each belt and between one belt and the next one. The noise level is therefore reduced.

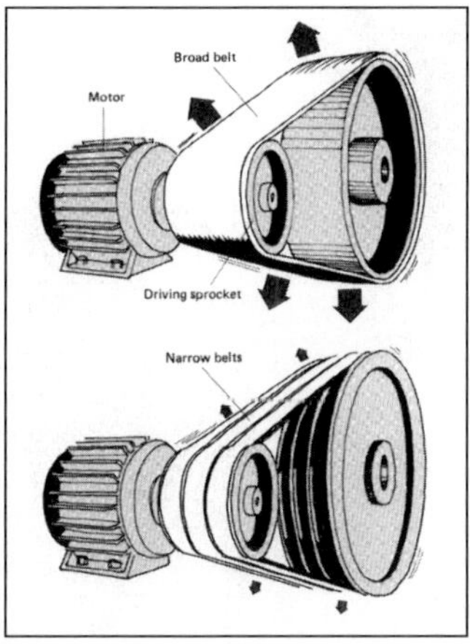

Example: Bins for the transport of material radiate noise when being loaded and emptied and when being transported over uneven surfaces. With this type of construction. pressure equalization can only occur around the upper edges.

Solution: The side panels can be fixed to edge frames with narrow brackets so that there is a much greater length of free edge around which pressure equalization can take place. If the size of the material or components allows. the side panels may be made of wire mesh to reduce radiation further

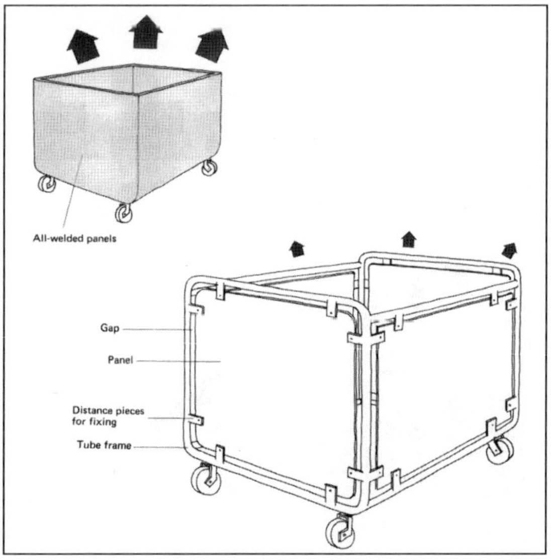

Damped Structures give Rise to Less Noise

If a panel is set into vibration, the level of vibration, and therefore the noise level will dimminish with time. The speed of this reduction depends on the material's internal damping. The higher the damping the quicker the drop in level. The damping also has an effect on the maximum level that can be generated from a given excitation; a well damped panel cannot be excited as much, as the resonances are reduced. Unfortunately most common metals have very low internal damping and a damping layer must usually be introduced in the form of a ready made laminate or as a spray or stick-on layer.

Principle:

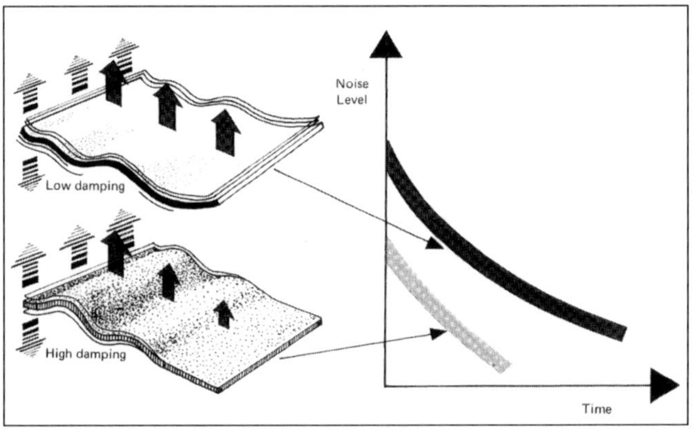

Example: Panels on machinery containing motors or pumps are prone to vibration and are therefore a normal source of radiated noise.

Solution: By using a laminated panel with high damping properties the noise can be reduced significantly.

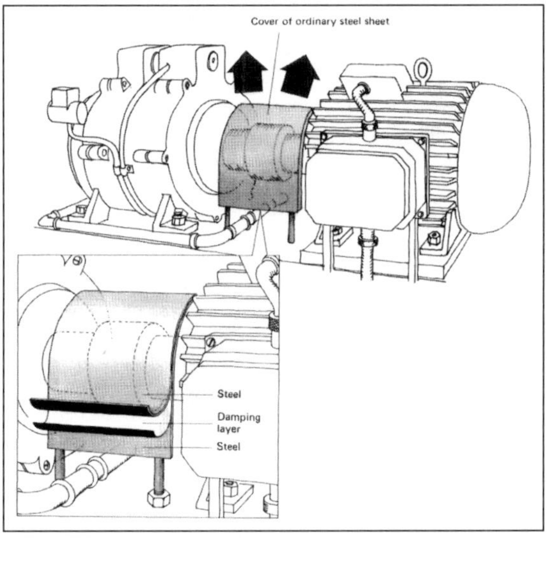

Resonances Amplify Noise Radiation but can be Damped Easily

Resonances strongly amplify the noise emitted by vibrating panel and plates, especially in homogenous structures. However, relatively small additions of extra damping can reduce the resonance peaks, and therefore the noise radiated, enormously. Pieces of damping material, fixed to a work piece temporarily, are also very effective.

Principle:

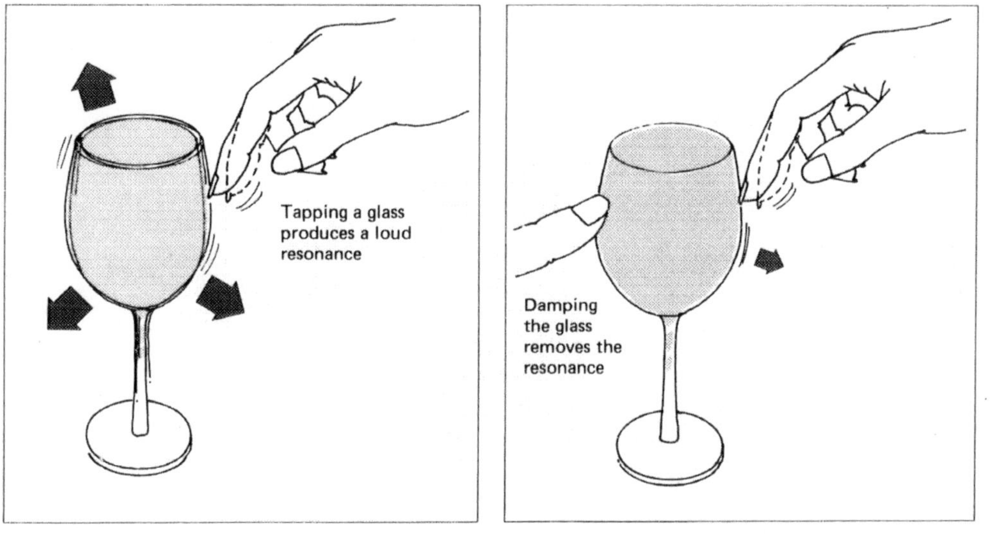

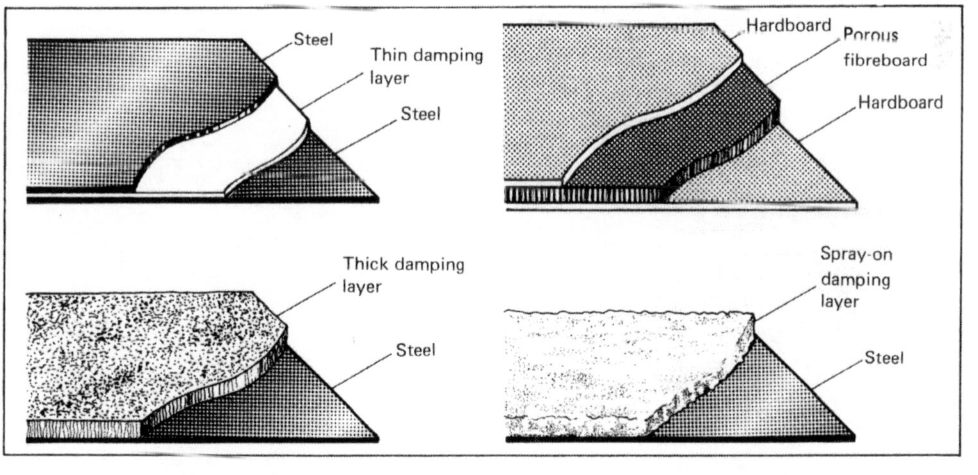

Example: A circular saw blade in a sharpening machine generates a high level of noise because of resonance and very low internal damping.

Solution: A disc of rubber damping material fastened to the blade by a stiff disc during sharpening, adds both mass and damping to the blade and reduces the amplification of the resonances.

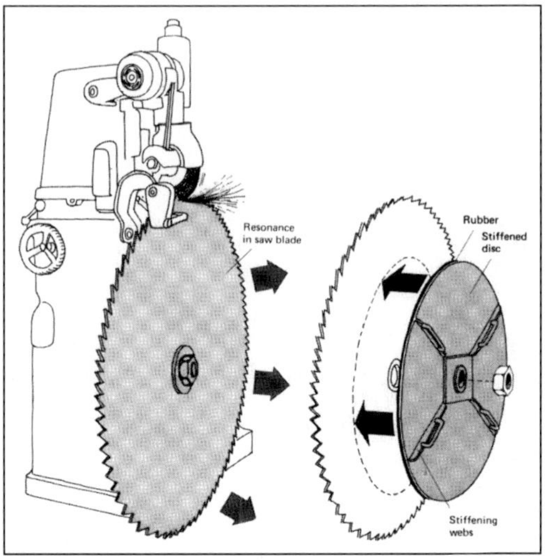

Example: During sawing operations on steel plate the sawing motion induces strong resonances in the work piece whose large area radiates a high level of screeching unpleasant noise.

Solution: Temporary addition of a magnetically held damping panel reduces the intensity of the resonances and reduces the noise to an acceptable level.

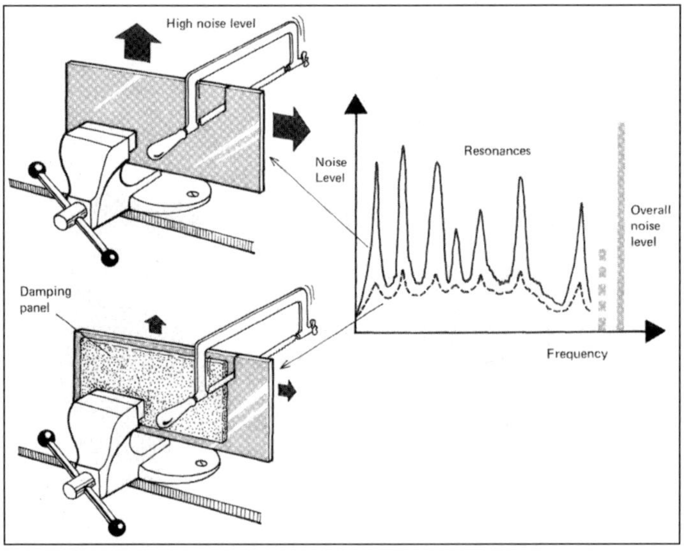

Example: The process of rivetting large structures such as aircraft, ship, or process plant components leads to high noise levels because of the impact caused and the large size of the component which efficiently converts the vibration energy into noise.

Solution: The application of temporary damping pads to the structure as it is rivetted reduces the intensity of the resonances and attenuates the vibration as it travels from the rivetting site to the rest of the panel.

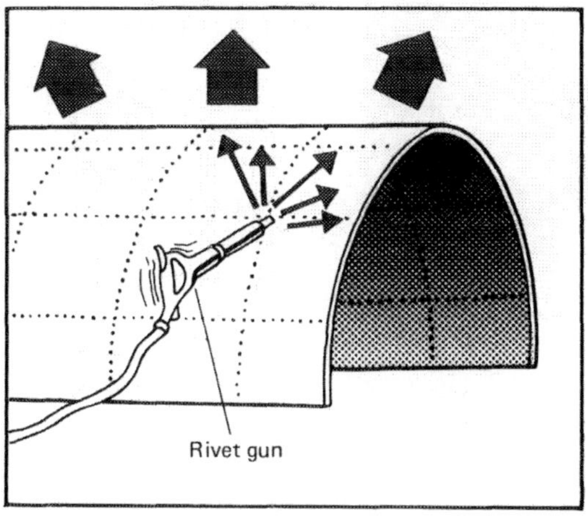

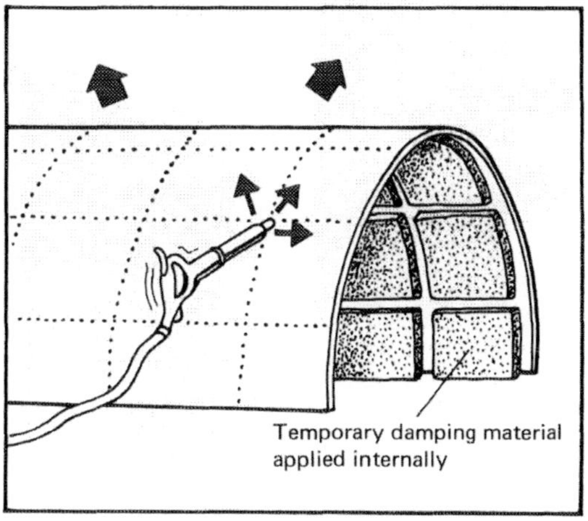

Resonances Transferred to a Higher Frequency are Easier to Damp

Large vibrating plates and shells often have low frequency resonances which are difficult to damp out. If the plate or shell can be stiffened, the resonance is shifted to a higher frequency which is easier to damp. In some cases it may still be difficult or expensive to apply the damping material to the separate areas, and therefore advantageous to mount a thin damped panel over the stiffening webs.

Principle:

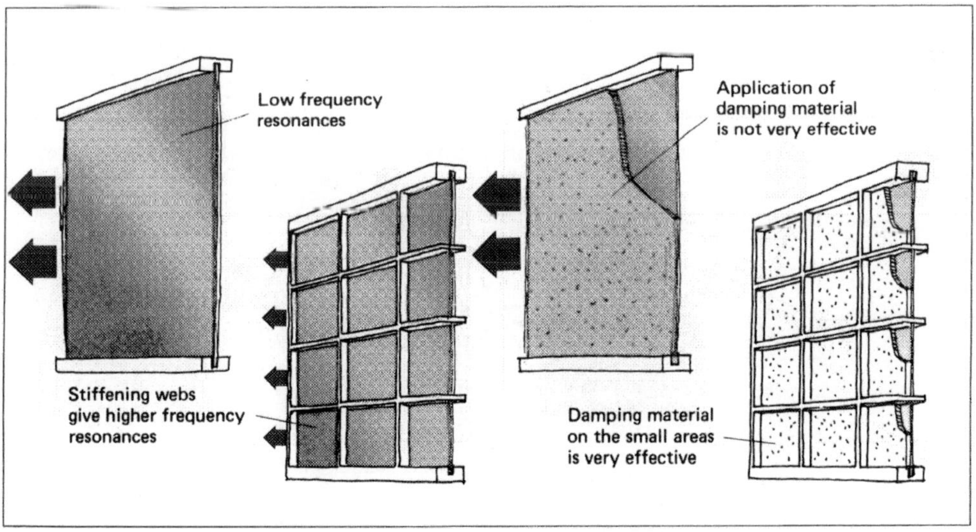

Example: The chief source of noise from a shearing machine proved to be radiation from the support and not from the workpiece as expected.

Solution: Stiffening webs were fitted to the support panels, and damped panels mounted on them.

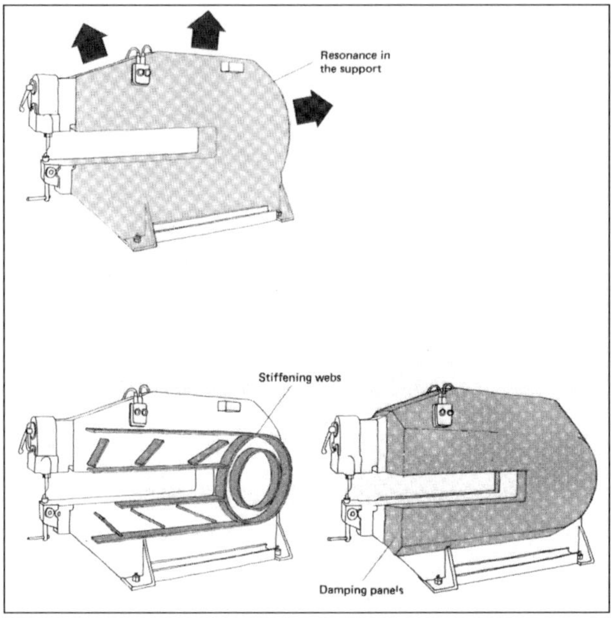

Flexible Mountings Isolate Machine Vibration

Nearly all structure-borne noise can be eliminated or at least significantly reduced by mounting vibration sources on flexible supports. In some cases it may be necessary to mount the receiver room on flexible mountings as well, e.g. where sensitive apparatus is used or for low level acoustic measurements. Normally it is preferable to isolate the source, or at least as near to the source as practically possible. In this way every region of the building is protected from structure-borne noise caused by the source. In many cases, especially in locations which are remote from the sound source, and therefore well insulated from the airborne sound produced by it, the structure borne noise is the most significant.

Principle:

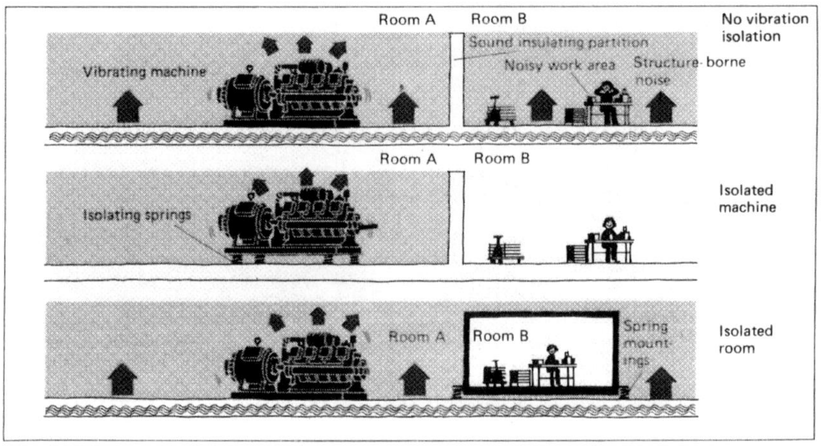

Example: Flexible mounts for vibration isolation can be obtained in a wide variety of types and materials to cope with any load requirement and any practical situation. For mounting heavy machinery, individual springs are used, with or without additional damping; for mounting light structures, pads of cork, expanded polystyrene, foam rubber, or rubber are often employed. Ceilings, ducts, and pipework are normally suspended from spring hangers or artificial rubber straps.

Incorrectly Chosen Mountings can Amplify Vibrations

A flexibly mounted machine always has a characteristic resonant frequency on its mountings. This resonance is determined by the weight of the machine and the stiffness of the mountings. A light machine and stiff mountings give a high resonance frequency; a heavy machine and low stiffness of the mountings give a low resonant frequency. Vibrations produced by the machine at frequencies lower than its mounted resonant frequency are not isolated. Vibrations well above the resonant frequency are isolated. Vibrations at the resonant frequency may be highly amplified if the internal damping of the mountings is low, and in any case will not be isolated. The natural frequency of the machine on its supports must always be below the normal running speed of the machine or the frequency of the vibrations to be isolated. Where the machine may spend some time at its resonant frequency, e.g. during run-up or run-down, mountings with very high internal damping should be chosen to keep the degree of amplification at resonance as low as possible.

Principle:

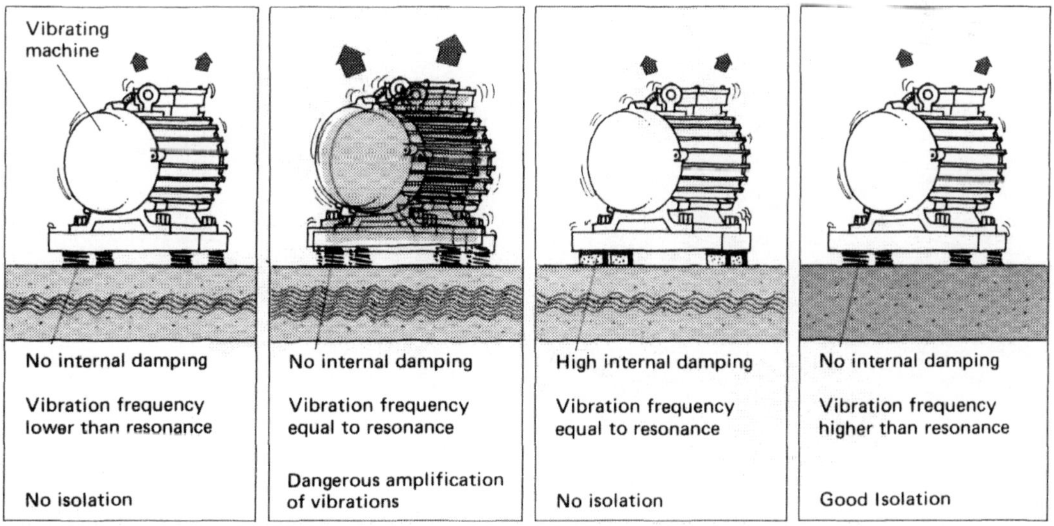

Example: Where a machine is run at a steady speed and continuously, without a large number of stops and starts, the resonant frequency of the suspension can be arranged to be well below the frequency of the machine and with a very low internal damping to give the highest isolation using, for example, simple spring isolators, If the machine spends a significant amount of time at resonance, e.g., because of frequent stops and starts of a compressor, this may lead to damage after a short time.

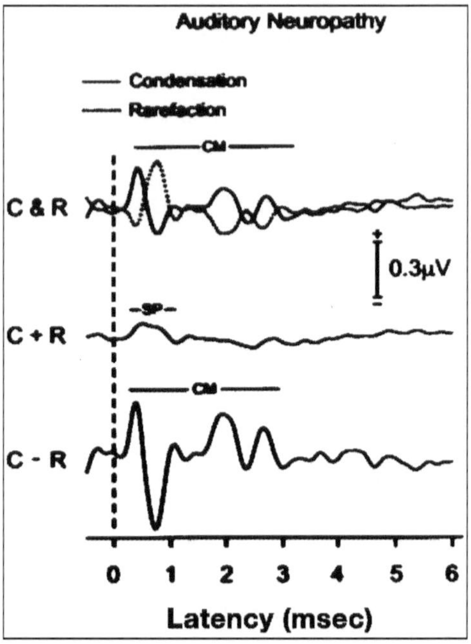

Solution: Supplying a type of isolator with high internal damping, e.g, pads of artificial rubber laminate, reduces the vibration significantly at the resonant frequency while only reducing the effectiveness of the isolation at the normal running speed by a small amount.

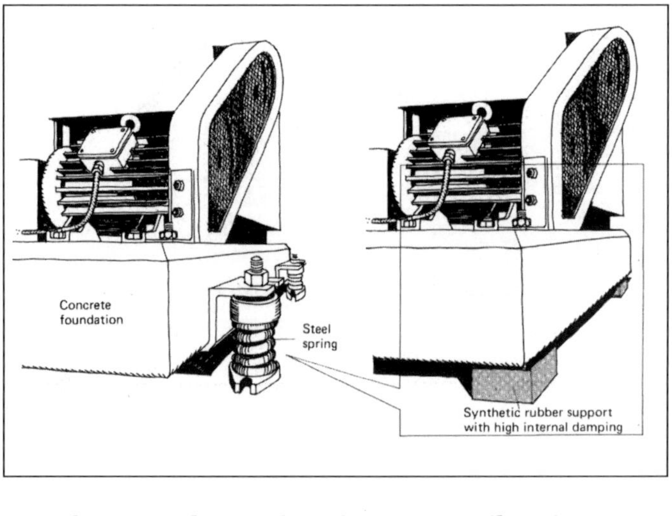

Example: A heavy machine producing low frequency vibration may cause the floor itself to resonate even though isolators of the correct rating are used. This problem is particularly common in concrete buildings, whose floors have low internal damping.

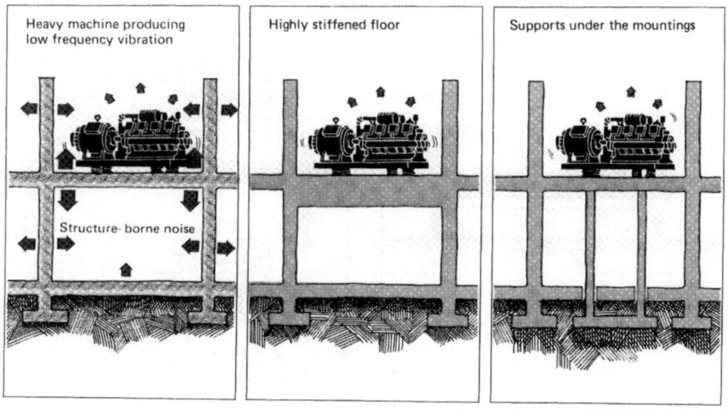

Solution: For the best isolation the natural frequency of the machine on its isolators should not only be well below the exciting frequencies from the machine, but should also be lower than the resonances of the floor. In practice, this may be achieved by reinforcing the floor structure to provide a more stiffer solid base. Alternatively, the machine may be mounted on pillars founded directly in the ground.

Displacement of all Mountings must be Equal to Avoid Rocking Motions

Flexible mountings for a machine should be chosen so that the static deflection at each mount is the same. If they are different, rocking motions may be forced at higher frequencies than the up-and-down motion of the machine, which are not effectively isolated by the mountings. On a machine with non-uniform weight distribution, mountings nearer the centre of gravity must be stiffer than those more remote from it.

The most efficient isolation is obtained if the mountings are fixed so that lines joining their points of application on the machine pass through the centre of gravity.

Principle:

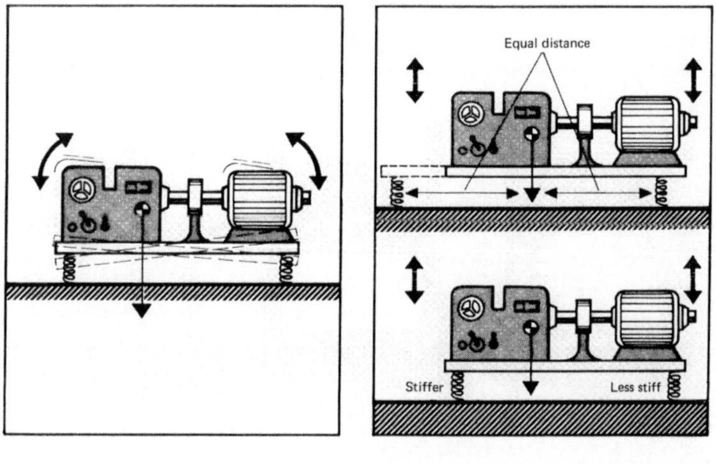

Example: If the centre of gravity of a vibrating machine is above the line of action of the horizontal forces from the mounting, there is the danger of setting up a rocking motion from this source. This is particularly a problem when dealing with out of balance forces on rotating machines where the sideways component is large.

Solution: Mounting the machine in a tray or on an inertia block enables the mounting points to be placed in the same horizontal plane as the centre of gravity.

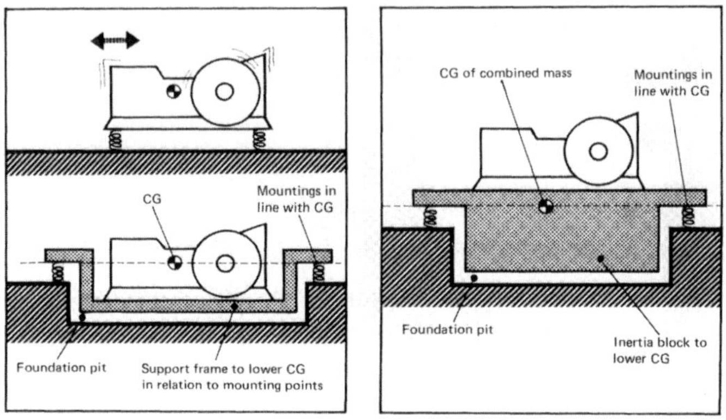

Structure-borne Sound via Connections must be Avoided

The most effective vibration isolation can be made totally ineffective if the vibration is transmitted by connections such as pipes, electrical conduits, supply ducts, etc. These must be flexible or contain flexible sections if vibration transmission is to be avoided.

Principle:

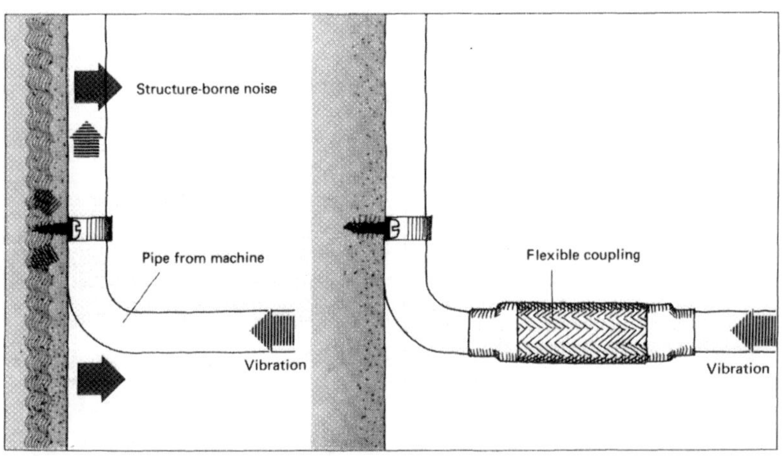

Example: Refrigeration plant can be a serious noise source because of the large pressure changes in the fluid during its passage through the compressor unit. Careful vibration isolation of the entire plant is necessary, and all ingoing and outgoing pipework should be isolated from the plant by flexible couplings.

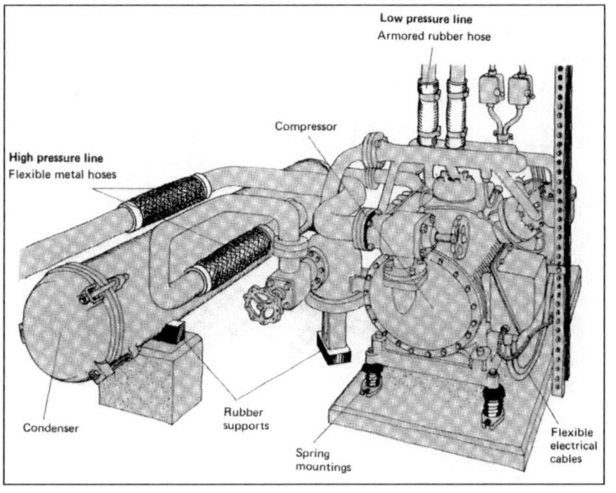

Insulation of Single Walls and Panels Depends on the Surface Density

When sound meets a wall, it sets the wall into vibration and sound is radiated from the other side. How much is radiated is a measure of the insulation of the wall, and is dependent on the surface density of the wall, i.e. its weight per unit area. In general, insulation increases as the frequency of the sound in~ creases or the thickness of the wall increases, until a point is reached where it begins to fall off again because of coincidence.

Example: What insulation does 15 mm chipboard give at 500 Hz? Surface weight is 10 kg/m^2. 10 x 500 = 5000. Insulation is 26 dB.

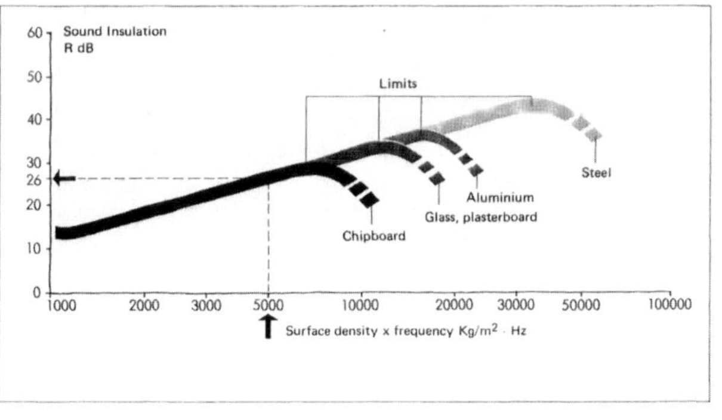

Example: A sand blasting system is the dominant noise source in a workshop, and is only separated from the rest of the area by thin curtains.

Solution: A sound insulated machine room can be built for the sand blasting plant and a partial enclosure erected round the work area. Access to this area is via heavy lead/ rubber laminate curtains which have high insulation while being flexible and easy to fold to allow easy access to the work area.

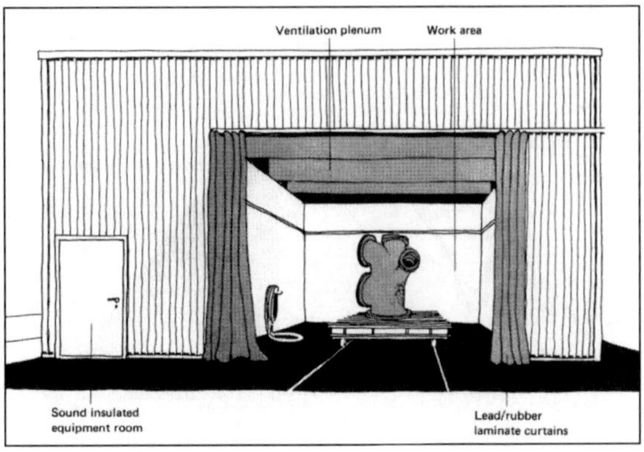

Single Walls have a Region of Poor Insulation

A single panel has natural bending modes which are dependent upon the stiffness and thickness of the panel. When the wavelengths of the incident sound and the panel modes are equal, they vibrate in sympathy and the panel's insulation ability is reduced before increasing again as the resonance condition is passed through. The coincidence dip only disappears if the internal damping of the panel is high. As the thickness of a single wall or panel is increased, its stiffness rises relatively faster than its weight and the region of low insulation comes down in frequency. At some frequencies it is possible for a thinner panel to give better insulation than one of greater thickness.

Principle:

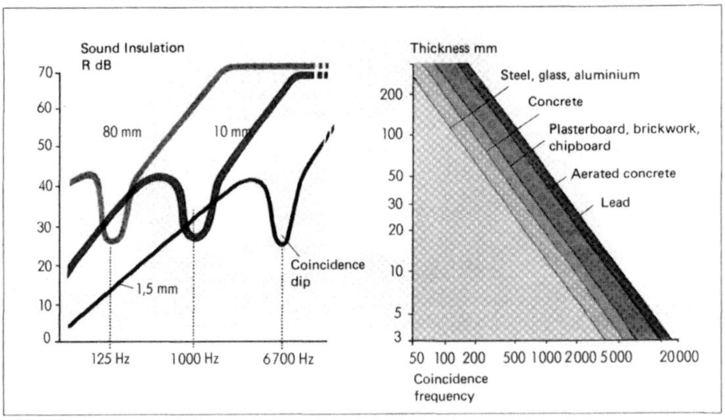

Example: One end of a workshop contained machines with a high noise output at 1000 Hz. The area had been separated from the rest of the workshop by a 25 mm chipboard partition with 6 mm windows. The insulation was not as expected because the partition had a coincidence dip at 1000 Hz, where the noise from the machines was also a maximum. The window's coincidence dip, on the other hand, was up at 2000 Hz.

Solution: Exchanging the partition with one made of two layers of plasterboard improved the situation by 10 dB. Although the weight was about the same, the new partition's stiffness was only a quarter of the previous one. The coincidence dip was therefore much higher at approximately 2500 Hz.

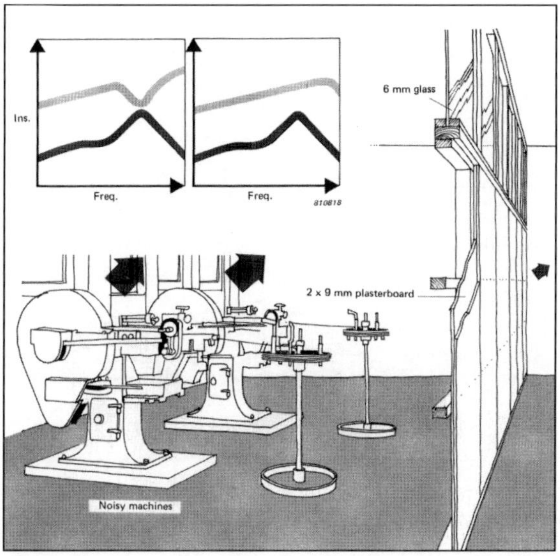

Stiffness and Weight are both Beneficial in Thick Walls

Most practical single walls have coincidence dips around 100 Hz. At frequencies above this, the insulation increases both with increased weight and increased stiffness. A cast concrete wall has a greater stiffness than a block one and may give the same insulation value though be lighter.

Principle:

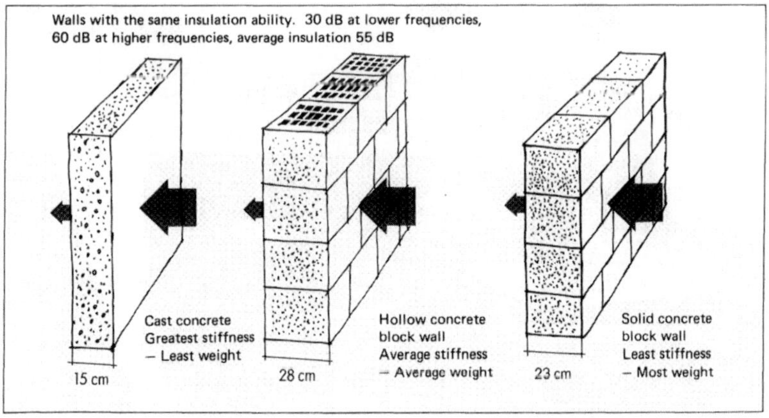

Example: The chief source of noise in an industrial building was found to be a fan unit.

Solution: A single block wall was built around the equipment with access via sound attenuating doors. This type of construction was chosen because it had to resist damage from transport trucks and also help to support the heavily loaded floor above.

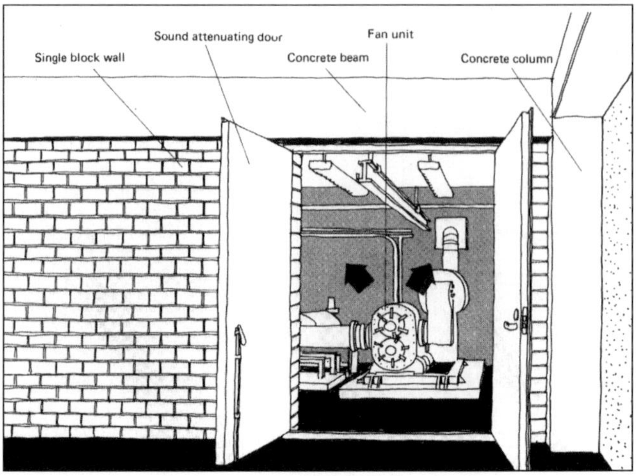

Lightweight Double Partitions give Good Insulation

Two light partitions separated by an air gap give an insulation which increases as the air gap increases. If an absorbent material is placed in the gap, the insulation is further increased. The most insulation is achieved if the two panels are completely disconnected from each other, mounted on separate frameworks. If connection is necessary the highest insulation is achieved if the ties are as small, as few, and as elastic as possible. Insulation can be obtained which would normally require a single partition of 5 or 10 times the weight.

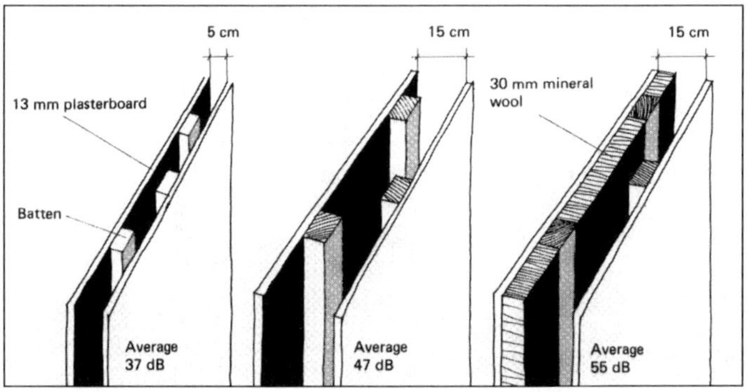

Example: Two connected workshops both contain noisy machinery, which is annoying in both locations.

Solution: The Noisy equipment should be located together at one end of one of the workshops and a double partition constructed with a large air gap in the form of an air lock, a gap large enough to serve as a corridor to the two quiet rooms. The doors are placed at opposite corners so that even if one of them is open an insulation of at least 35 dB is achieved. When closed over 50 dB can be expected. To achieve the same insulation with a single wall would require heavy masonary or concrete construction and the use of a special sealed sound-proof door.

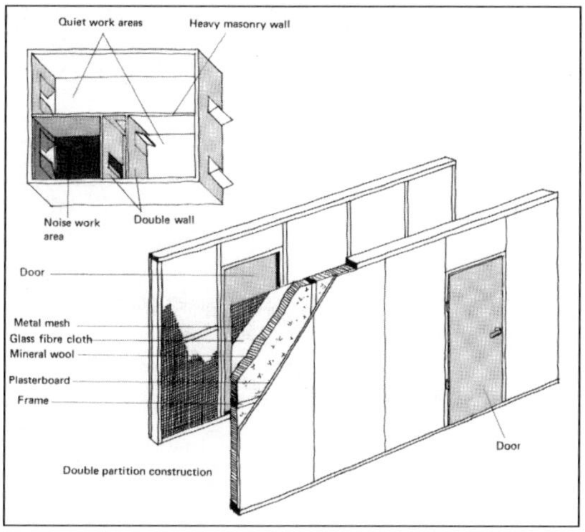

Thick Porous Layers are Effective Absorbers at both High and Low Frequencies

Porous materials through which air can pass are often excellent sound absorbers. Some examples are: Glass fibre, mineral wool, foam rubber, woodwool and sintered metal. If the material has closed cells the absorption is low. Thin layers are only capable of absorbing high frequencies, whereas thick layers can absorb over a wide frequency range including both high and low frequencies. To be effective below about 100 Hz a layer has to be impractically thick or mounted over an air space.

Principle:

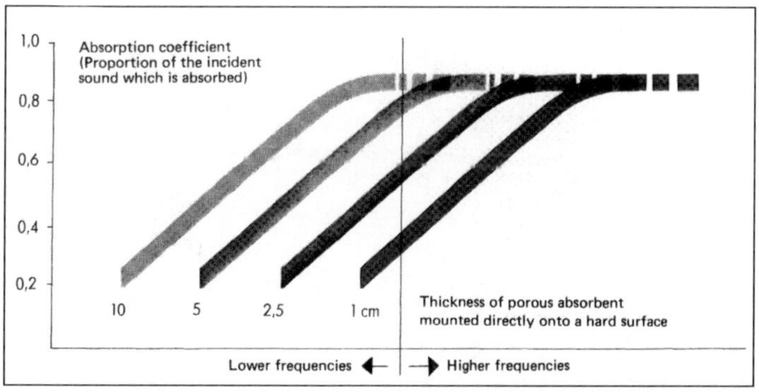

Example: A workshop with a high noise level, especially at low frequencies, has to be treated to reduce the noise levels over the entire frequency range. Hanging panel absorbers can be used in a large part of the workshop where the ceiling is free of obstructions. These are very efficient, having two absorbing sides to each panel. A traversing crane makes it impossible to use these in the other part of the workshop. Instead, horizontal absorbent panels are mounted well below the ceiling to obtain improved low frequency absorption. Except in regions close to a noise source it is possible to reduce the overall noise level by up to 10 dB.

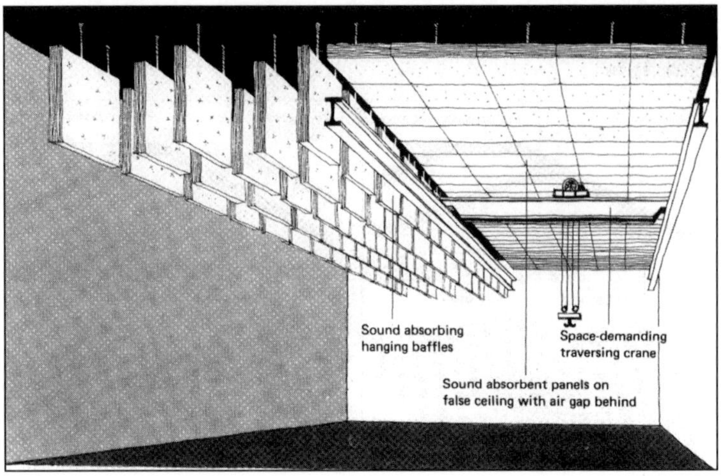

Perforated Panels over Absorbent need not Reduce its Effectiveness

In order to protect the absorbent and improve its appearance it is often covered with a perforated material of some kind. This does not significantly alter the characteristics of the absorbent if the perforations are a sufficient proportion of the total area. A perforation ratio of 15% is sufficient for thin panels. The ratio must be greater, the thicker the panel becomes. In addition, it is generally better to perforate with many small holes than with a smaller number of large holse.

Principle:

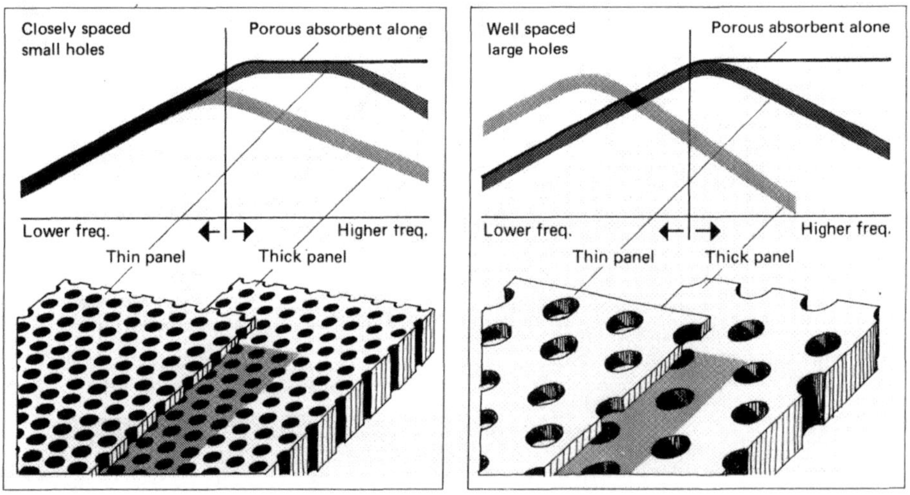

Example: In order to improve the appearance in a building it is possible to choose from a wide variety of materials to cover large areas of absorbent. The only requirement is that the uncovered area is sufficiently large to allow the absorbent material to do its job satisfactorily. Textiles, wood strips, expanded metal, and various types of partially open panel can be used to create the desired appearance.

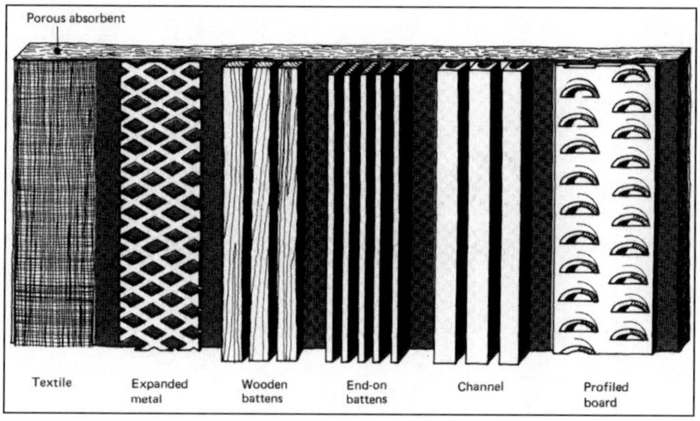

Panel Absorbers are Effective at Low Frequencies

Thin panels mounted on a framework absorb low frequencies well in a fairly narrow range, whose frequency depends on the size and thickness of the panel and its distance from the wall. The effective absorption bandwidth depends on the internal damping of the panel; high internal damping giving a wider frequency range. For a porous absorbent to be as good at low frequencies, it would have to be extremely thick.

Principle:

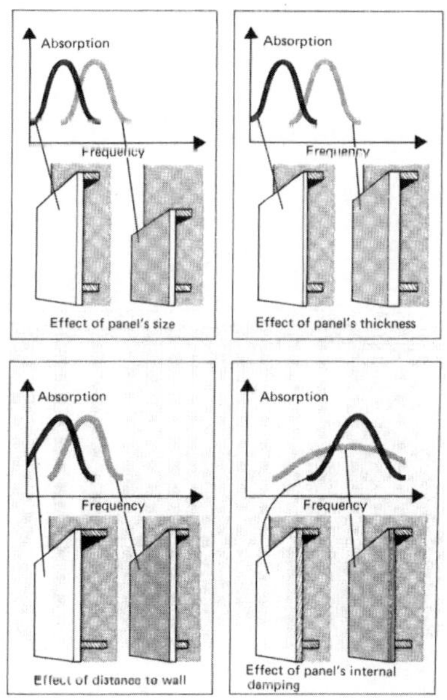

Example: In an engine test room a resonance is excited when the test engine is run at or near its normal speed. At speeds away from the normal running speed the resonance disappears completely.

Solution: The walls may be covered in panels fixed to a wooden frame, the dimensions being chosen so that frequency range includes the unwanted resonance. To make the panel effective at frequencies at either side of the resonance, a panel material with high internal damping should be chosen, e.g. a laminate or fibre board.

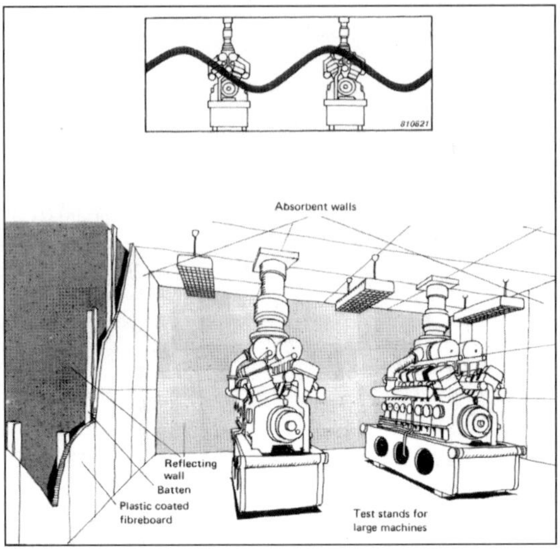

Screens should be Combined with Absorbent Ceilings

Where the noise is dominated by high frequencies, a noise screen can be very effective against it. The screen is more effective the higher it is and the nearer to the source it is placed. However, if the ceiling above the source is non-absorbent, it reflects noise from the source into the "quiet area" on the far side of the screen. An absorbent ceiling prevents this reflection taking place, and improves the performance of the screen.

Principle:

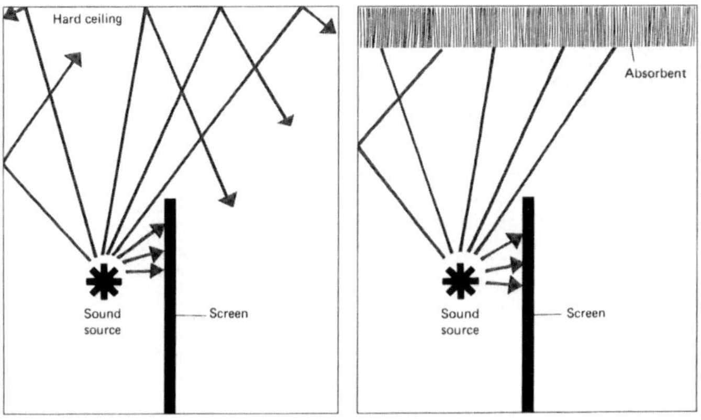

Example: In an assembly hall with a number of production lines running parallel. One of the lines produces much more noise than the others. Noise from body preparation produces high frequency noise which affects everyone in the assembly hall.

Solution: By using lightweight absorbent screens on both sides of the noisy line, and hanging sound absorbing baffles above it, the noise levels are reduced at the quieter lines while not worsening the situation for those actually working on the noisy line. If it had merely been enclosed in a sound attenuating tunnel without sound absorbents, the reverberent sound level would have increased locally.

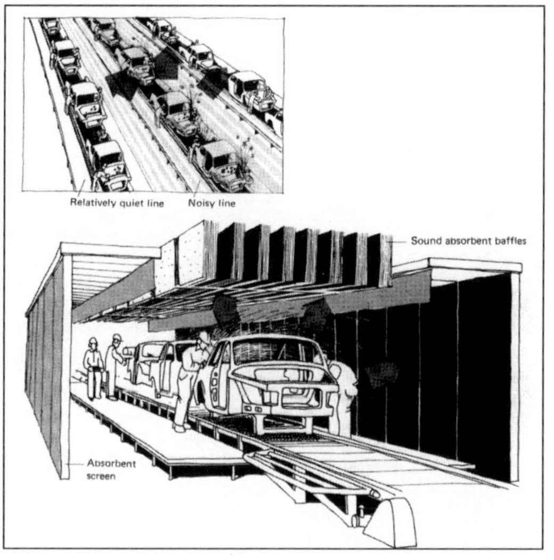

Changes in Ducts Reduce Noise Transmission

At all changes along the transmission path, part of the sound energy is reflected back toward the source. In a duct this change may be the shape or area of the cross-section, bends, branches and wall material. This fact can often be exploited during design of a duct system to achieve a level of attenuation. Attenuation obtained in this way is termed reactive.

Principle:

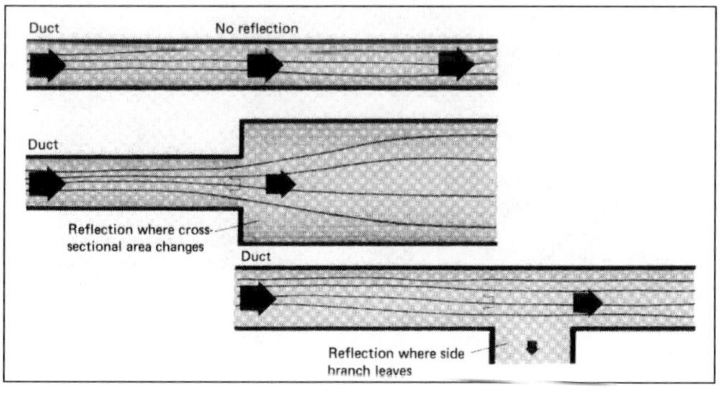

Example: A large landscaped office is to be fitted with a mechanical ventilation system. Space limitations prevent the use of an attenuator on the fan outlet large enough to provide the necessary noise reduction.

Solution: By making use of the sound reduction obtained when the duct contains a number of area changes, direction changes, and distribution branches, it is possible to achieve the required sound level limits at the delivery vents, even though the fan could not be sufficiently silenced at source. Absorbent on the duct walls and smooth bends avoid the generation of more noise and a better distribution of air is achieved around the office.

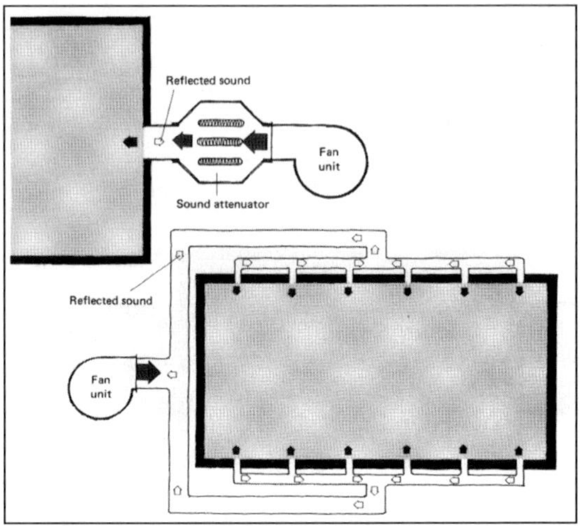

Reactive Attenuators are Efficient in a Narrow Frequency Range

A reactive attenuator is an effective method of reducing low frequency noise over a limited frequency range, and is relatively compact. By coupling a number of attenuators of different sizes together, most conveniently within the same external casing, it is possible to cover an extended range of frequencies. Perforated tubes are often used within the attenuator to improve gas flow and provide some absorption.

Principle:

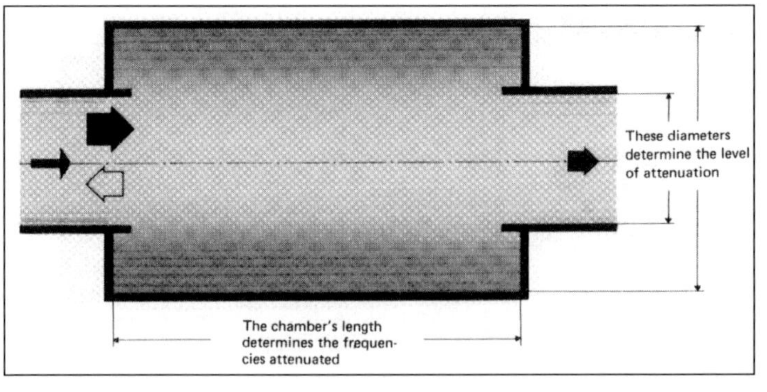

Example: Absorbent (resistive) attenuators are simple to design and produce, and provide noise reduction over a wide frequency range. However, the absorbent material can be easily blocked with residues and carbon deposits which make it less effective after a time, and absorption of combustible materials can cause a fire risk.

Solution: A multi-step reactive attenuator can be used in this case. It attenuates over a wide range of frequencies, is less sensitive to deposits, and rugged. The example shown is for a large piston engine, and is especially suitable where the motor speed only varies over a fairly narrow range.

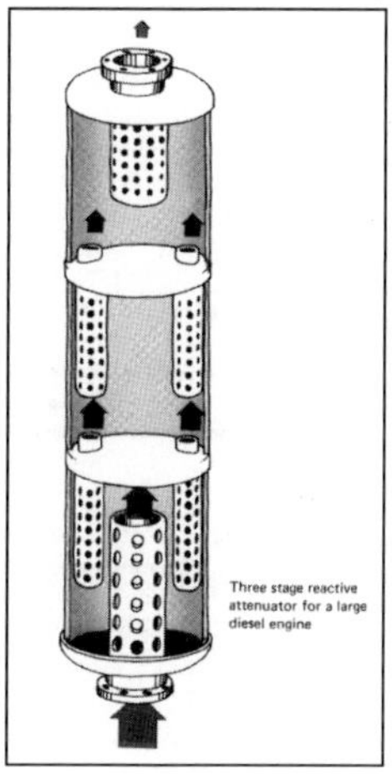

Expansion Chambers are Effective where Low Frequencies Dominate

If a pipe or duct is provided with an enlargement or chamber, then the low frequency fluctuations are evened out. This is very effective where continuous pressure pulses are concerned, e.g. in engine exhausts or compressor outlets. The lower the frequency, the larger the chamber must be to be effective.

Principle:

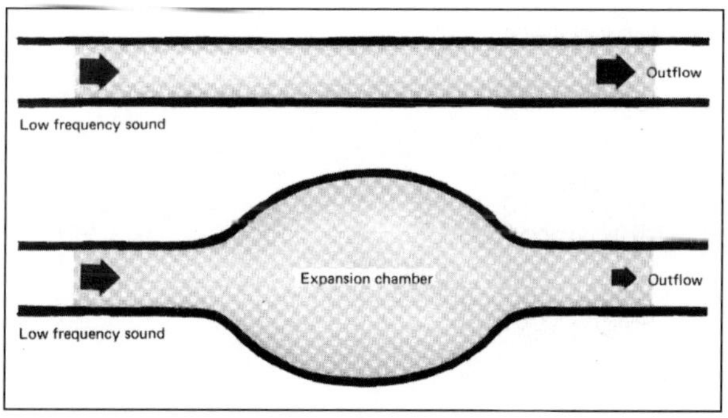

Example: The exhaust of a pneumatic drill produces both high and low frequency noise; low frequency from the repetitive pulses, and high frequency from the jet of gas escaping with each pulse.

Solution: By surrounding the drill body with a jacket acting as an expansion chamber for the low frequencies, and as a noise screen for the high frequencies, the noise level is reduced substantially. Because the energy in the pulse is spread more evenly over a longer time by the jacket. The noise is also reduced in that way. As the impact of the drill on the concrete is an inherently noisy process which cannot be easily quietened, there is a limit to the amount of quietening which is worthwhile.

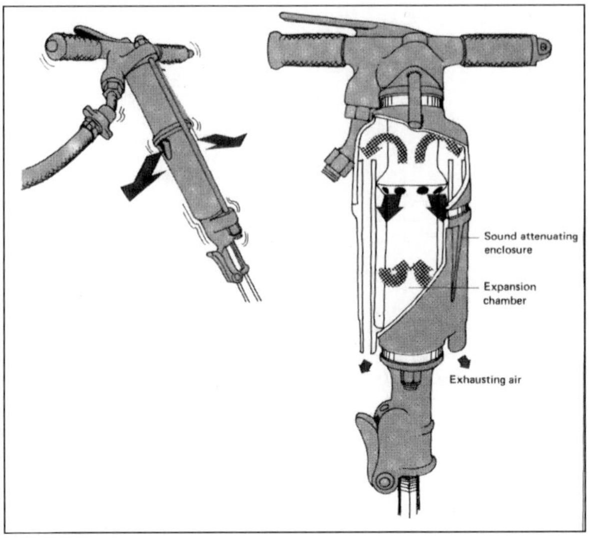

Pure Tones can be Reduced by Interference

When a noise contains a predominant tone or a few tones, it can be substantially reduced by interference, i.e. the presence of two out of phase tones of the same frequency at the same place, which tend to cancel each other out. An interference attenuator consists of a branch which leaves and later returns to the main channel. Sound passing through it travels an odd number of half wavelengths further than in the main channel, so that when the two sounds meet again they are out of phase and cancel each other out.

Principle:

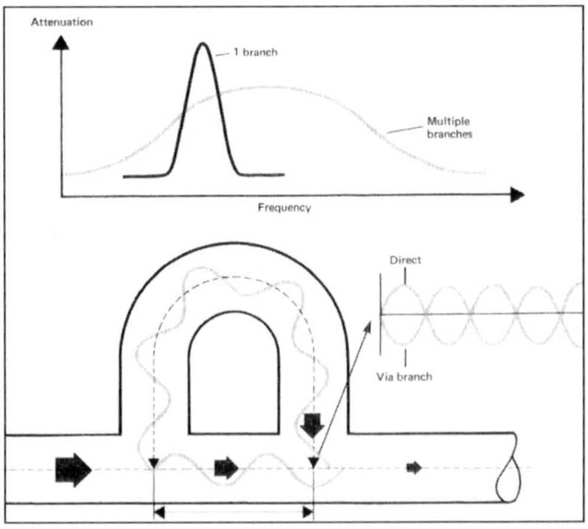

Example: If a tone is not completely fixed, the bandwidth of the interference attenuator can be broadened by having a number of branches of slightly different lengths. The attenuation at a given frequency will be a little lower, however. This type of device is suitable for engines operating at a more or less constant speed, generator sets, for example, and for fan and blower units.

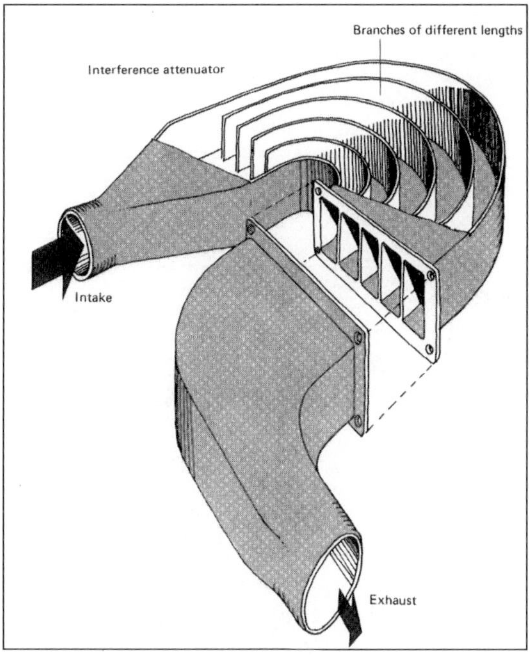

Unused Spaces can be used as Absorbent Plenum Chambers

An absorbent room is a simple but effective sound attenuator. Ducts can be fed into a chamber whose walls are lined with sound absorbent material which absorbs the sound energy. To prevent the direct passage of high frequencies, the incoming and outgoing ducts should not be directly opposite each other. The larger the volume of the chamber and the thicker the absorbent lining, the lower the frequencies which can be effectively absorbed.

Principle:

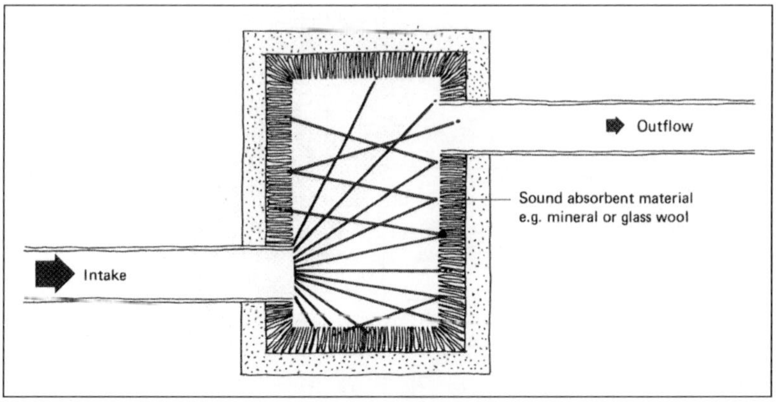

Example: The shape of the space used as a plenum chamber is not important, as long as it is well lined with absorbent and large enough to deal with the lowest frequencies encountered, so any unused part of the building can be utilised. It should also be well sealed in order to prevent leakage of the air out of the system.

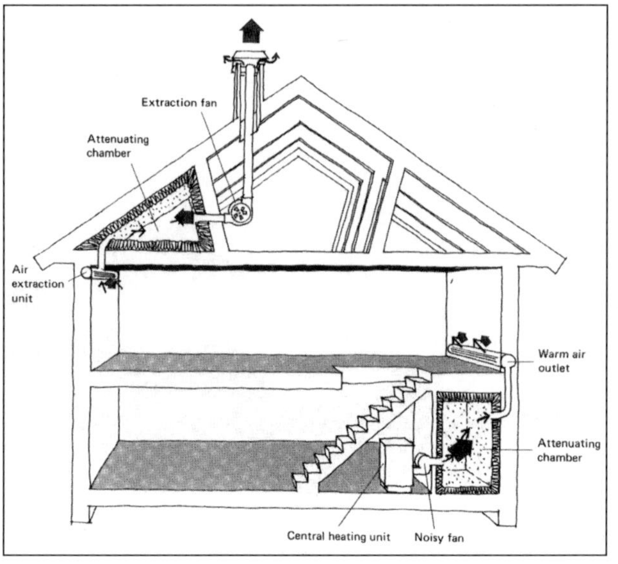

Absorbent Attenuators are Effective over a Wide Frequency Range

The simplest type of absorbent attenuator is a duct with sound absorbent material on the walls. The thicker the materials, the lower the frequency which can be absorbed. For higher frequencies though, thinner absorbent layers are effective, but the large gap allows noise to pass directly along. This layers and narrow passages are therefore more effective at high frequencies. For good absorption over the widest frequency range, thick absorbent layers and narrow passages are best.

Principle:

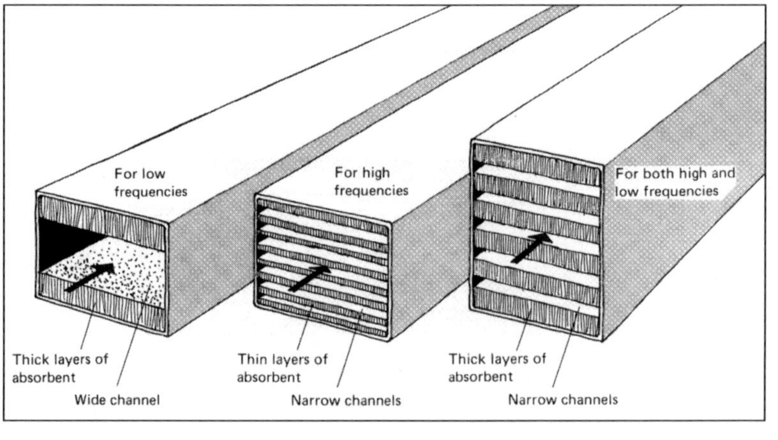

Example: Gas turbine powered standby generator sets are used extensively both as emergency power supplies and to complement normal generating plant in peak periods. It can be necessary to quieten the set by anything up to 70 dB, over a wide frequency range. Extensive use is made of absorbent materials in the form of splitters, baffles, and linings on the walls of plenum chambers on both the intake and exhaust sides.

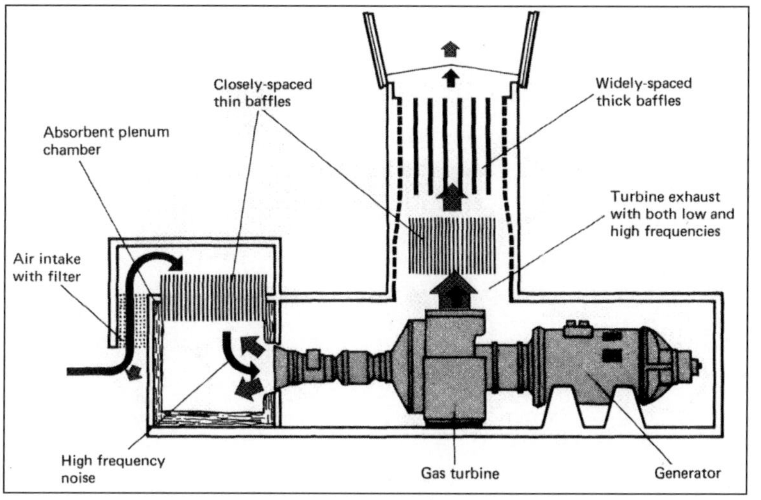

Wind Generated Tones can be Avoided by Profile Changes or Spoilers

When air flows past an object, a powerful pure tone, known as a Karman tone, can be produced at certain wind speeds. The tone is caused by the regular shedding of vortices from alternate edges on the downwind side of the object as the laminar boundary layer separates from the surface. In addition to a loud tone this phenomenon may also lead to severe structural vibration and damage if the modes of vibration of the object are at the same frequency. By lengthening the object in the direction of the air flow, i.e. by "streamlining", it is possible to keep the boundary layer attached right round the object, greatly reducing the tone level. Where the flow may come from any direction this method is not practicable, e.g. where chimneys are concerned. Spoilers can then be added to break up the air flow so that random turbulence is produced which does not generate the disturbing pure tone.

Principle:

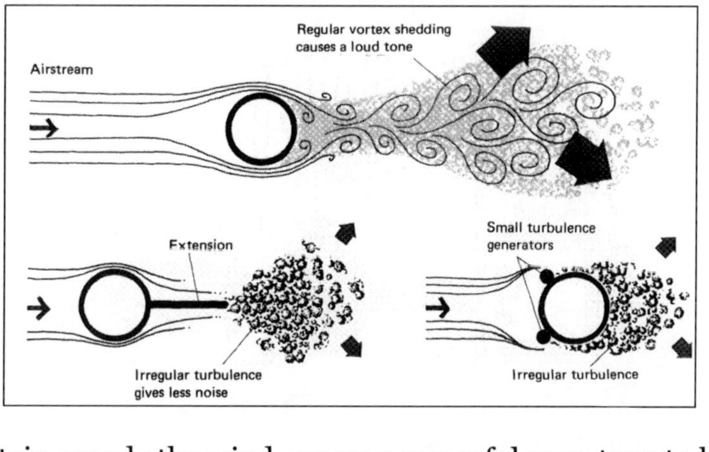

Example: At certain speeds the wind causes a powerful pure tone to be emitted from the chimney.

Solution: A spiral fin of metal mounted on the chimney causes local tubulence regardless of the wind direction, so regular vortex shedding and the tone which arises from it are avoided. The pitch of the spiral should not be the same all the way up. As well as presenting an irregular shape to a wind coming from any direction, the spiral also increases the strength and stiffness of the chimney.

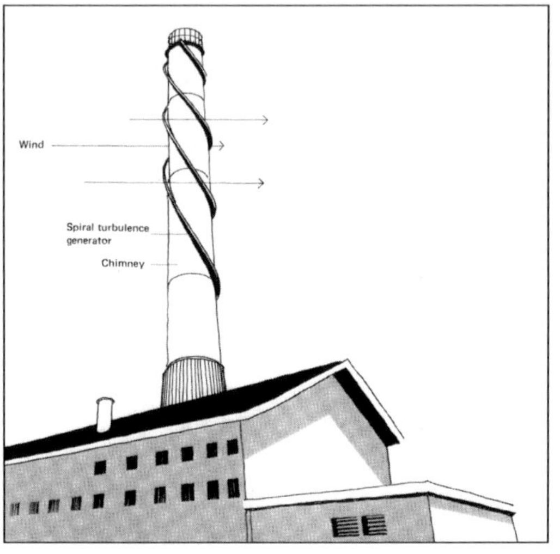

NOISE CONTROL

Noise control is an active or passive means of reducing sound emissions, often incentivised by personal comfort, environmental considerations or legal compliance.

Practical and efficient noise control is wholly reliant on an accurate diagnosis of what is causing the noise, which first involves finding the source of noise.

Once the source of noise has been found, the focus is reducing the noise at source by engineering means. (The most common noise sources can be divided into aerodynamic (fans, pneumatics, combustion, etc.) and mechanical (impacts, friction, etc.) Effective noise control focuses on reducing the noise from these sources as close to the source as possible. Noise control for aerodynamic sources include quiet air nozzles, pneumatic silencers and quiet fan technology.

In architectural acoustics and environmental acoustics, noise control refers to the set of practices employed for noise mitigation. Within architectural acoustics these practices include:

"Interior sound reverberation reduction, inter-room noise transfer mitigation and exterior building skin augmentation. More specific architectural noise control methods include the installation of acoustical gypsum, ceiling tiles, ceiling panels, carpet and draperies. In the field of environmental sound, common noise control practices include: design of noise barriers, development and enforcement of noise abatement legal codes and urban design."

There are four basic principles of noise control:

Sound Insulation

Prevent the transmission of noise by the introduction of a mass barrier. Common materials have high-density properties such as brick, concrete, metal etc.

Sound Absorption

A porous material which acts as a 'noise sponge' by converting the sound energy into heat within the material. Common sound absorption materials include open cell foams and fiberglass.

Vibration Damping

Applicable for large vibrating surfaces. The damping mechanism works by extracting the vibration energy from the thin sheet and dissipating it as heat. A common material is sound deadened steel.

Vibration Isolation

Prevents transmission of vibration energy from a source to a receiver by introducing a flexible element or a physical break. Common vibration isolators are springs, rubber mounts, cork etc.

The noise pollution can be controlled at the source of generation itself by employing techniques like:

- Reducing the noise levels from domestic sectors;
- Maintenance of automobiles;
- Control over vibrations;
- Low voice speaking;
- Prohibition on usage of loud speakers;
- Selection of machinery;
- Maintenance of machines;
- Control in the transmission path;

- Installation of barriers;
- Installation of panels or enclosures;
- Groundproofing is the process of developing acoustic treatments to suppress.

Soundproofing

Soundproofing is the process of developing acoustic treatments to suppress, reflect, diffuse or absorb sound waves either at the source using products such as baffling or insulation on automotive engines, pipes/ducting or machinery rooms, or at the destination using technologies such as double glazed windows and insulation to reduce unwanted external noise from road traffic, airports and other rooms within offices, apartments and other residential dwellings.

Soundproofing Techniques

Sound Absorption

Porous or resonant absorbing materials can be installed to convert sound wave energy into a tiny amount of heat effectively reducing the amount of sound that is reflected around a room. Soundblock Solutions use Plano Bevel Tiles for sound absorption within restaurants, hotels, schools, home theatres and many other commercial and residential applications. Soundblock Solutions carry a range of sound absorbing products including: Echohush Profile Panels, Sonex Acoustic Foam, Quietstone Light and Reapor Soundproofing.

Sound Diffusion

Irregular surfaces can be used to break down, or scatter sound waves so that they travel along many smaller paths. This results in the overall energy of the sound wave depleting faster and therefore travelling less distance.

Sound Reflection

Reflecting sound waves is a common technique for reducing traffic noise on freeways and other major traffic routes. Typically hard surfaces such as concrete or glass are used to bounce sound waves upwards (into the sky) or away from buildings. The overall architecture of a building including material and angles can also be used to effectively reflect sound.

Soundproofing Categories

Reducing Sound within a Room

Soundproof rooms, also known as anechoic (echo free) or semi-anechoic chambers, are designed to reduce unwanted reverberation and echo within a room for the purposes of sound-quality (e.g. recording studios and home theatres) or noise mitigation e.g. hotels, bars, restaurants, and open-plan offices. Room soundproofing typically also includes insulation to reduce unwanted external noise.

Limiting Sound Leakage

Soundproofing materials can be utilised to limit the transference of sound waves between rooms (either adjacent, above or below) and from external sources. A range of items within a room or rooms can be soundproofed including windows (double glazing), floors, walls, ceilings and doors. Existing or traditional windows can be fitted with an acoustic secondary windows using one of our two retrofitted Soundblock solutions; Soundout Aluminium Window or Soundblock Acrylic Magnetic Window.

RADIO NOISE REDUCTION SYSTEMS

Dolby Noise-reduction System

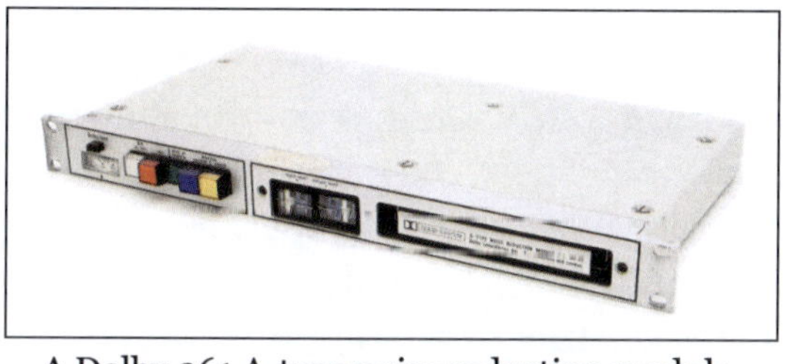

A Dolby 361 A-type noise reduction module.

A Dolby noise-reduction system, or Dolby NR, is one of a series of noise reduction systems developed by Dolby Laboratories for use in analog audio tape recording. The first was *Dolby A*, a professional broadband noise reduction for recording studios in 1965, but the best-known is *Dolby B*, a sliding band system for the consumer market, which helped make high fidelity practical on cassette tapes which used a relatively noisy tape size and speed. It is common on high fidelity stereo tape players and recorders to the present day. Of the noise reduction systems, *Dolby A* and *Dolby SR* were developed for professional use. *Dolby B*, *C*, and *S* were designed for the consumer market. Aside from Dolby IIX, all the Dolby variants work by companding, or compressing the dynamic range of the sound during recording and ex*panding* it during playback.

Usage

When recording a signal on magnetic tape, there is a low level of "noise" in the background which sounds like hissing. One solution is "low-noise" tape which records more signal, and less "noise". Other solutions are to run tape at a higher speed or use a wider tape. Cassette tapes were originally designed to trade off fidelity for convenience of recording voice by using a very narrow tape run at a very slow speed of 1 7/8 inches per second (ips) in a simple plastic shell when 15 or 7.5 ips was for high fidelity and 3 3/4 ips was lower fidelity. When Advent designed one of the first cassette tape decks for high fidelity, noise reduction was one way to fix problems with the format, along with chrome and high-bias tapes with extended high frequency response. Within 10 to 20 years, the original reel-to-reel high fidelity tape recorders which did not need noise reduction were eventually replaced by cassette decks for home use. The dominant noise reduction scheme "Dolby B" was widely accepted because if an inexpensive cassette player lacked the switch, they would just sound brighter which often offset the dull sounds of cheap players.

The signal to noise ratio is simply how large the music signal is compared to the low level of the "noise" with no signal. When the music is loud, the low hiss is not noticeable, but when the music is soft or in silence, most of what can be heard is the noise. If the recording level is adjusted so that the music is always loud, then it could in theory be turned down later, and the noise volume would also be turned down. The idea is for electronics to automatically increase the recording volume when it is soft, but reduce the volume on playback. Some schemes like Dolby B concentrate only on the high frequencies so that the "hiss" sound of noise will be masked when volume is turned down for playback.

Dolby noise reduction is a form of dynamic preemphasis employed during recording, plus a form of dynamic deemphasis used during playback, that work in tandem to improve the signal-to-noise ratio. While Dolby A operates across the whole spectrum, the other systems specifically emphasize the audible frequency range where background tape hiss, an artifact of the recording process that is similar to white noise, is most noticeable (usually above 1 kHz, or two octaves above Middle C).

The Dolby preemphasis boosts the recorded level of the quieter audio signal at these higher frequencies during recording, effectively compressing the dynamic range of that portion of the signal, so that quieter sounds above 1 kHz receive a proportionally greater boost. As the tape is recorded, the relative amplitude of the signal above 1 kHz is used to determine how much pre-emphasis to apply - a low-level signal is boosted by 10 dB (Dolby B) or 20 dB (Dolby C). As the signal rises in amplitude, less and less pre-emphasis is applied until at the "Dolby level" (0 VU), no signal modification is performed.

The sound is thus recorded at a higher overall level on the tape relative to the tape's overall noise level, requiring the tape formulation to preserve this specially recorded signal without distortion. On playback, the opposite process is applied (deemphasis), based on the relative signal component above 1 kHz. Thus as this portion of the signal decreases in amplitude, the higher frequencies are progressively more sharply attenuated, which also filters out the constant background noise on the tape when and where it would be most noticeable.

The two (pre and de-emphasis) processes are intended to cancel each other out as far as the actual recorded program is concerned. Only de-emphasis is applied to the incoming signal and noise during playback. After playback de-emphasis is complete, apparent noise in the output signal is reduced, and this process should not produce any effect noticeable to the listener (other than reduced noise of course). Playback without noise reduction produces a noticeably brighter sound, however.

The calibration of the recording and playback circuitry is therefore critical for faithful reproduction of the original program content, and this is easily offset by poor quality tape, dirty recording/playback heads, or using inappropriate bias levels/frequency for the tape formulation, as well as tape speed, when recording or duplicating. This can manifest itself as muffled-sounding playback, or "breathing" of the noise level as the signal varies.

On some high end consumer equipment, Dolby calibration control is included: for recording, a reference tone at Dolby level may be recorded for accurate playback level calibration on another transport; at playback, the same recorded tone should produce the identical output, as indicated by a Dolby logo marking at 0 VU on the VU meter(s). (In consumer equipment Dolby Level is defined as 200 nWb/m; calibration tapes were available to assist correct level setting.) For accurate off-the-tape monitoring during recording on 3-head decks, both processes must be employed at once, and circuitry provided to accomplish this is marketed under the rubric "Double Dolby".

Dolby A

Dolby *A* was the company's first noise reduction system, presented in 1965. It was intended for use in professional recording studios, where it became commonplace, gaining widespread acceptance at the same time that multitrack recording became standard. The input signal is split into frequency bands by four filters with 12 dB per octave slopes, with cutoff frequencies (3 dB down points) as follows: high–pass at 80 Hz; band–pass from 80 Hz to 3 kHz; a low–pass from 3 kHz; and another low–pass at 9 kHz. (The stacking of contributions from the two low-pass bands allows greater noise reduction in the upper frequencies.) The compander circuit has a threshold of -40 dB, with a ratio of 2:1 for a compression/expansion of 10 dB. This provides about 10 dB of noise reduction increasing to a possible 15 dB at 15 kHz, according to articles written by Ray Dolby in JAES and Audio.

As with the "B" system, correct matching of the compression and expansion processes is important. The calibration of the expansion (decoding) unit for magnetic tape uses a flux level of 185 nWb/m, which is the level used on industry calibration tapes such as those from Ampex; this is set to 0 VU on the tape recorder playback and to Dolby Level on the noise reduction unit. In the record (compression or encoding) mode a characteristic tone (Dolby Tone) generated inside the noise reduction unit is set to 0 VU on the tape recorder and to 185 nWb/m on the tape.

Dolby A also saw some use as the method of noise reduction in optical sound for motion pictures.

Dolby B

A typical front loading consumer autoreverse hi-fi cassette deck from the late 1980s, featuring full electronic transport, Dolby B and C noise reduction, and HXPro dynamic headroom expansion.

Dolby B was developed after Dolby A and presented in 1968, as a single sliding band system providing about 9 dB noise reduction (A-weighted), primarily for cassettes. It was much simpler than Dolby A and therefore much less expensive to implement in consumer products. Dolby B recordings are acceptable when played back on equipment that does not possess a Dolby B decoder, such as most inexpensive portable and car cassette players. Without the de-emphasis of the decoder, the sound will be perceived as brighter as high frequencies are emphasized, which can be used to offset “dull” high frequency response in inexpensive equipment. However, Dolby B provides less effective noise reduction than Dolby A, generally by a factor of more than 3 dB.

From the mid-1970s, Dolby *B* became standard on commercially pre-recorded music cassettes in spite of the fact that some low-end equipment lacked decoding circuitry, although it allows for acceptable playback on such equipment. Most pre-recorded cassettes use this variant. VHS video recorders used Dolby B on linear stereo audio tracks.

Up to 1980 tape recorders supporting Dolby *B* were often just labelled Dolby *NR*.

Dolby FM

In the early 1970s, some expected Dolby NR to become normal in FM radio broadcasts and some tuners and amplifiers were manufactured with decoding circuitry; there were also some tape recorders with a Dolby B “pass-through” mode. In 1971 WFMT started to transmit programs with Dolby NR, and soon some 17 stations broadcast with noise reduction, but by 1974 it was already on the decline. Dolby FM was based on Dolby B, but used a modified 25 μs pre-emphasis time constant and a frequency selective companding arrangement to reduce noise.

A similar system named High Com FM was tested in Germany between July 1979 and December 1981 by IRT. It was based on the Telefunken High Com broadband compander system, but never introduced commercially in FM broadcasting.

Dolby C

Dolby *C* was introduced in 1980. It provides about 15 dB noise reduction (A-weighted). It is constructed by combining the effect of two Dolby B systems together with an expansion to lower frequencies. The resulting recordings sound much worse when played back on equipment that does not have Dolby *C* noise reduction. Some of this harshness can be mitigated by using Dolby B on playback. Dolby C first appeared on higher end cassette decks in the 1980s. The first commercially available cassette deck with Dolby C was NAD 6150C, which came into the market in ca. 1981. It was also used on professional video equipment for the audio tracks of the Betacam and Umatic SP videocassette formats. In Japan the first cassette deck with Dolby C was the AD-FF5 from Aiwa.

Dolby SR

The Dolby *SR* (Spectral Recording) system, introduced in 1986, was the company's second professional noise reduction system. It is a much more aggressive noise reduction approach than Dolby *A*. It attempts to maximise the recorded signal at all times using a complex series of filters that change according to the input signal. As a result, Dolby *SR* is much more expensive to implement than Dolby *B* or *C*, but Dolby *SR* is capable of providing up to 25 dB noise reduction in the high frequency range. It is only found on professional recording equipment.

In the motion picture industry, as far as it concerns distribution prints of movies, the Dolby A and SR markings refer to Dolby Surround which is not just a method of noise reduction, but more importantly encodes two additional audio channels on the standard optical soundtrack, giving left, center, right, and surround.

SR prints are fairly well backward compatible with old Dolby A equipment. The Dolby SR-D marking refers to both analog Dolby SR and digital Dolby Digital soundtracks on one print.

Dolby S

Dolby *S* was introduced in 1989. It was intended that Dolby *S* would become standard on commercial pre-recorded music cassettes in much the same way that Dolby *B* had in the 1970s, but it came to market when the Compact Cassette was being replaced by the Compact Disc as the dominant mass market music format. Dolby Labs claimed that most members of the general public could not differentiate between the sound of a CD and a Dolby *S* encoded cassette. Dolby *S* mostly appeared on high-end audio equipment and was never widely used.

Dolby *S* is much more resistant to playback problems caused by noise from the tape transport mechanism than Dolby C. Likewise, Dolby S was also claimed to have playback compatibility with Dolby B in that a Dolby S recording could be played back on older Dolby B equipment with some benefit being realized. It is basically a cut down version of Dolby *SR* and uses many of the same noise reduction techniques. Dolby *S* is capable of 10 dB of noise reduction at low frequencies and up to 24 dB of noise reduction at high frequencies.

Dolby HX/HX-Pro

Dolby HX-Pro was invented in 1980 and patented in 1981 (EP 0046410) by Jørgen Selmer Jensen of Bang & Olufsen.

B&O immediately licensed HX-Pro to Dolby Laboratories, stipulating a priority period of several years for use in consumer products, to protect their own Beocord 9000 cassette tape deck.

Magnetic tape is inherently non-linear in nature due to hysteresis of the magnetic material. If an analogue signal were recorded directly onto magnetic tape, its reproduction would be extremely distorted due to this non-linearity.

To overcome this, a high frequency signal, known as bias, is mixed in with the recorded signal, which "pushes" the envelope of the signal into the linear region.

If the audio signal contains strong high frequency content, in particular from percussion instruments such as a high-hat, this adds to the constant bias causing magnetic saturation on the tape. Dolby HX Pro automatically reduces the bias signal in the presence of strong high frequency signals, making it possible to record at a higher signal level, leading to its name: HX = Headroom eXtension.

HX-Pro only applies during recording; the improved signal to noise ratio is available no matter which tape deck the tape is played back on, and therefore HX-Pro is not a noise-reduction system in the same way as Dolby A, B, C & S, although it does help to improve noise reduction encode/decode tracking accuracy by reducing tape performance non-linearity.

Some record companies issued HX-Pro pre-recorded cassette tapes during the late 1980s and early 1990s.

Technological Trends

The widespread proliferation of digital audio in professional and consumer applications (e.g. compact discs, music download) has made analogue audio production less prevalent, but Dolby's analogue noise reduction systems are still widely used in niche analogue production environments.

FMX (Broadcasting)

FMX is the name of a commercially unsuccessful noise reduction system developed in the 1980s for FM broadcasting in the United States.

FMX Logo.

FM stereo broadcasting is known to incur up to a 23 dB noise penalty over that of monophonic FM broadcasting; this is due to the combination of the triangular FM noise spectrum and the wider baseband bandwidth occupied by the stereo multiplex signal. Developed at the CBS Technology Center, FMX was intended to improve this characteristic for listeners in the fringe areas where the noise penalty would be worst. This improvement was achieved by adding an amplitude-compressed version of the L–R (left-minus-right, or difference) signal modulated in quadrature with the stereo subcarrier, using a version of the CX noise-reduction system originally developed at CBS for LP records.

Upon his accession as Chairman of CBS, Laurence Tisch closed the CBS labs in 1986, whereupon the FMX intellectual property was spun off to an investment group, under the name Broadcast Technology Partners (BTP).

With about 50 stations using the technology and perhaps another 50 committed, a controversy emerged in 1989 when MIT professor and Bose Corporation CEO Amar Bose and Bose engineer William Short released a critical study, finding the system to have the potential to "seriously degrade the quality of stereo reception whether received by FMX equipment or not." According to the study, the heavy compression of the L–R audio caused interference in receivers under multipath conditions. A BTP spokesman rebutted the finding, saying that the Bose analysis exaggerated a "worst-case scenario", and actually employed "flawed mathematics" to attempt to prove their point. Perhaps not coincidentally, a patent application filed earlier by Bose disclosed an invention whose implementation was incompatible with the adoption of FMX, while another – filed by Bose and Short around the time of the release of the critical study – sought to improve the performance of FMX under multipath conditions.

Despite industrial supporters in both the broadcast and consumer electronics industries, the system never achieved a critical mass, and faded into obscurity. However, numerous related patents have since been applied for or granted, referencing the original FMX patents.

ACTIVE NOISE CONTROL

Active noise control (ANC), also known as noise cancellation, or active noise reduction (ANR), is a method for reducing unwanted sound by the addition of a second sound specifically designed to cancel the first.

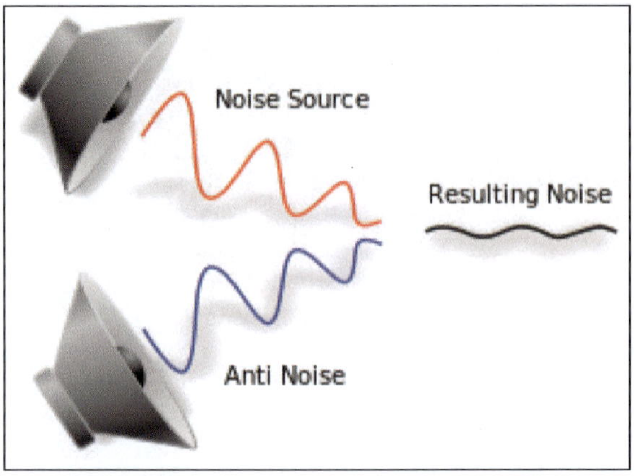

Graphical depiction of active noise reduction.

Sound is a pressure wave, which consists of alternating periods of compression and rarefaction. A noise-cancellation speaker emits a sound wave with the same amplitude but with inverted phase (also known as antiphase) to the original sound. The waves combine to form a new wave, in a process called interference, and effectively cancel each other out – an effect which is called destructive interference.

Modern active noise control is generally achieved through the use of analog circuits or digital signal processing. Adaptive algorithms are designed to analyze the waveform of the background aural or nonaural noise, then based on the specific algorithm generate a signal that will either phase shift or invert the polarity of the original signal. This inverted signal (in antiphase) is then amplified and a transducer creates a sound wave directly proportional to the amplitude of the original waveform, creating destructive interference. This effectively reduces the volume of the perceivable noise.

A noise-cancellation speaker may be co-located with the sound source to be attenuated. In this case it must have the same audio power level as the source of the unwanted sound. Alternatively, the transducer emitting the cancellation signal may be located at the location where sound attenuation is wanted (e.g. the user's ear). This requires a much lower power level for cancellation but is effective only for a single user. Noise cancellation at other locations is more difficult as the three-dimensional wavefronts of the unwanted sound and the cancellation signal could match and create alternating zones of constructive and destructive interference, reducing noise in some spots while doubling noise in others. In small enclosed spaces (e.g. the passenger compartment of a car) global noise reduction can be achieved via multiple speakers and feedback microphones, and measurement of the modal responses of the enclosure.

Applications

Applications can be "1-dimensional" or 3-dimensional, depending on the type of zone to protect. Periodic sounds, even complex ones, are easier to cancel than random sounds due to the repetition in the wave form.

Protection of a "1-dimension zone" is easier and requires only one or two microphones and speakers to be effective. Several commercial applications have been successful: noise-cancelling headphones, active mufflers, anti-snoring devices, vocal or center channel extraction for karaoke machines, and the control of noise in air conditioning ducts. The term "1-dimension" refers to a simple pistonic relationship between the noise and the active speaker (mechanical noise reduction) or between the active speaker and the listener (headphones).

Protection of a 3-dimension zone requires many microphones and speakers, making it more expensive. Noise reduction is more easily achieved with a single listener remaining stationary but if there are multiple listeners or if the single listener turns his head or moves throughout the space then the noise reduction challenge is made much more difficult. High frequency waves are difficult to reduce in three dimensions due to their relatively short audio wavelength in air. The wavelength in air of sinusoidal noise at approximately 800 Hz is double the distance of the average person's left ear to the right ear; such a noise coming directly from the front will be easily reduced by an active system but coming from the side will tend to cancel at one ear while being reinforced at the other, making the noise louder, not softer. High frequency sounds above 1000 Hz tend to cancel and reinforce unpredictably from many directions. In some, the most effective noise reduction in three-dimensional space involves low frequency sounds. Commercial applications of 3-D noise reduction include the protection of aircraft cabins and car interiors, but in these situations, protection is mainly limited to the cancellation of repetitive (or periodic) noise such as engine-, propeller- or rotor-induced noise. This is because an engine's cyclic nature makes fast Fourier transform analysis and the noise cancellation easier to apply.

Modern mobile phones use a multi-microphone design to cancel out ambient noise from the speech signal. Software is used to subtract the sound captured by the microphones placed furthest from the mouth, from the microphone closest to the mouth, which produces a much clearer speech quality.

Active vs. Passive Noise Control

Noise control is an active or passive means of reducing sound emissions, often for personal comfort, environmental considerations or legal compliance. *Active* noise control is sound reduction using a power source. *Passive* noise control is sound reduction by noise-isolating materials such as insulation, sound-absorbing tiles, or a muffler rather than a power source.

Active noise canceling is best suited for low frequencies. For higher frequencies, the spacing requirements for free space and zone of silence techniques become prohibitive. In acoustic cavity and duct based systems, the number of nodes grows rapidly with increasing frequency, which quickly makes active noise control techniques unmanageable. Passive treatments become more effective at higher frequencies and often provide an adequate solution without the need for active control.

ACOUSTIBLOK

Acoustiblok is a registered trademark product of Acoustiblok Inc. It is a sound attenuation material that was introduced in 1999 and is manufactured in the United States. It is typically used during new construction or remodeling projects and is designed for application beneath the drywall and other finishing materials.

Physical Properties

Acoustiblok is a mass-loaded vinyl, sound isolation product. It is an extremely dense but very flexible material that comes in two sizes; 16oz. and 32oz:

- Sixteen ounce Acoustiblok is 1/8" thick, and weighs 1 lb/ft^2.
- Thirty-two ounce Acoustiblok is 1/4" thick, and weighs 2 lb/ft^2.

Performance Data

The ability of a partition to block transmission of airborne sound is typically described by a single number rating of the Sound Transmission Class (STC):

- Sixteen ounce Acoustiblok material has an STC of 26.
- Thirty-two ounce Acoustiblok material has an STC of 32.
- Adding one layer of Acoustiblok to a standard metal stud gypsum board interior wall, increases its STC to 53.
- Double steel stud partition walls, faced with single layers of drywall and containing fiberglass insulation and Acoustiblok, can achieve an STC of 66.

STC is calculated using a range of frequencies similar to the human voice; attenuation in higher and lower ranges should be estimated from the Sound Transmission Loss graphs provided with standard STC reports.

Test procedures and the formula for calculation of STC values are defined in ASTM Standards E90-02 and E413-87.

Construction Applications

New Construction and Remodel/Renovation

Acoustiblok material is typically installed in partition walls and floor/ceiling assemblies when they are stripped to the studs.

Classification

Acoustiblok is qualified for use in commercial and institutional construction by its UL

Classification file, found under Fire Tests of Building Construction and Materials. This Classification includes use in wall designs of the U300, U400, and V400 series, and floor-ceiling constructions of the L500 series, as specified in ANSI/UL 263 Fire Resistance Ratings. Ratings of 1 and 2 hours can be obtained, depending on specific partition design.

ACOUSTIC QUIETING

Acoustic quieting is the process of making machinery quieter by damping vibrations to prevent them from reaching the observer. Machinery vibrates, causing sound waves in air, hydroacoustic waves in water, and mechanical stresses in solid matter. Quieting is achieved by absorbing the vibrational energy or minimizing the source of the vibration. It may also be redirected away from the observer.

One of the major reasons for the development of acoustic quieting techniques was for making submarines difficult to detect by sonar. This military goal of the mid- and late-twentieth century allowed the technology to be adapted to many industries and products, such as computers (e.g. hard drive technology), automobiles (e.g. motor mounts), and even sporting goods (e.g. golf clubs).

Aspects of Acoustic Quieting

When the goal is acoustic quieting, a number of different aspects might be considered. Each aspect of acoustics can be taken alone or in concert so that the end result is that the reception of noise by the observer is minimized.

Acoustic quieting might consider.

- Noise generation: By limiting the noise at its source,
- Sympathetic vibrations: By acoustic decoupling,
- Resonations: By acoustic damping or changing the size of the resonator,
- Sound transmissions: By reducing transmission using many methods (depending whether the transmission is through air, liquid, or solid), or
- Sound reflections: By limiting the reflection using many methods, e.g. by using acoustic absorption (deadening) materials, trapping the sound, opening a "window" to let sound out, etc.

By analyzing the entire sequence of events, from the source to the observer, an acoustic engineer can provide many ways to quieten the machine. The challenge is to do this in a practical and inexpensive way. The engineer might focus on changing materials, using a damping material, isolating the machine, running the machine in a vacuum, or running the machine slower.

Methods of Quieting

Mechanical Acoustic Quieting

- Sound isolation: Noise isolation is isolating noise to prevent it from transferring out of one area, using barriers like deadening materials to trap sound and vibrational energy. Example: In home and office construction, many builders place sound-control barriers (such as fiberglass batting) in walls to deaden the transmission of noise through them.

A sound proof room, showing acoustic damping tiles used for noise absorption and soundproofing.

- Noise absorption: In architectural acoustics, unwanted sounds can be absorbed rather than reflected inside the room of an observer. This is useful for noises with no point source and when a listener needs to hear sounds only from a point source and not echo reflections. Example: In a recording studio, sound proofing is accomplished with bass traps and anechoic chambers. Wallace Sabine, an American physicist, is credited with studying sound reverberations in 1900, and Carl Eyring revised his equations in 1930 for Bell Labs. Another example is the ubiquitous use of dropped ceilings and acoustical tiles in modern office buildings with high ceilings. Submarine hulls have special coatings that absorb sound.
- Acoustic damping: Vibration isolation prevents vibration from transferring beyond the device into another material. Damping mounts have progressed in the industry to offer vibrational resistance in many degrees of freedom. Recent advances include shock isolators damping in at least six degrees of freedom. Acoustic damping also has uses in seismic shock protection of buildings. Motors and rotating shafts are commonly fitted with these mounts at the points where they contact the building or the chassis of a large machine.
- Acoustic decoupling: Certain parts of a machine can be built to keep the frame, chassis, or external shafts from receiving unwanted vibrations from a moving

part. Example: Volkswagen has registered a patent for an "acoustically decoupled underbody for a motor vehicle." Another example: Western Digital has registered a patent for an "acoustic vibration decoupler for a disk drive pivot bearing assembly."

- Preventing stalls: Whenever a machine undergoes an aerodynamic stall, it will abruptly vibrate.
- Preventing cavitation: When a machine is in contact with a fluid, it may be susceptible to cavitation. The sounds of gas bubbles imploding is the source of the noise. Ships and submarines which have screws that *cavitate* are more vulnerable to detection by sonar.
- Preventing water hammer: In hydraulics and plumbing, water hammer is a known cause for the failure of piping systems. It also generates considerable noise. A valve that abruptly opens or shuts is the most common cause for water hammer.
- Shock absorption: Just as automotive shock absorbers are used to prevent mechanical shocks from reaching the passengers in a car, they are also important for quieting shocks.
- Reduction of resonance: Essentially any piece of metal or glass has certain frequencies to which it is susceptible to resonate. A machine that resonates would make a tremendous noise. Resonance also occurs in enclosures, such as when echoes reverberate in an ocarina or the pipe of a pipe organ.
- Material selection: By choosing nonmetallic components, the transmission of sound and vibrations can be minimized. For example: instead of using rigid brass fittings, a machine using flexible plastic pipe fittings may be much quieter. In some cases air can be evacuated from a machine and sealed hermetically, the vacuum inside becoming a barrier to sound transmission. In cases where porous plastic materials are used in acoustic applications, the porosity of the plastic is adjusted to either dampen specific wavelengths or for minimal sound loss in a speaker grill cover.

Quieting for Specific Observers

- Underwater acoustics: All of the above types of acoustic quieting apply to submarines. Additionally, a submarine may employ a tactic that prevent sounds from reaching a listener at a particular ocean depth. Operating below the depth of the sound channel axis, where the speed of sound in water is the lowest, a submarine can prevent detection by surface ships.
- Sound refraction: Just as a submarine can use refraction to hide its acoustic signature from surface vessels, the same principle of sound refraction can be used

to prevent certain observers from hearing the noise. For example, an outdoor observer close to the ground will have sound waves refracted toward him when the ground is *cooler* than the ambient air and away from him when the ground is *hotter* than the air.

- Sound redirection: One of the obvious ways to reduce the received sound level of an observer is to place the observer out of the path of the highest amplitude sounds. For example, if we mark off a circle around a jet engine and make sound power level observations along that circle, we would expect that the sound is loudest directly in line with the jet's exhaust. Observations perpendicular to the exhaust would be significantly quieter.
- Hearing protection: An observer may be forced to wear ear plugs in areas of high ambient noise levels. This may be the only quieting method available in areas of noise pollution, such as an open-air firing range or an airport.

Electronic Quieting

- Electronic vibration control: Electronics, sensors, and computers are now employed to reduce vibration. Using high speed logic, vibrations can be damped quickly and effectively by counteracting the motion before it exceeds a certain threshold.
- Electronic noise control: Electronics, sensors, and computers are also employed to cancel noise by using phase cancellation which matches the sound amplitude with a wave of the opposite polarity. This method employs the use of an active sound generating device, such as a loudspeaker to counteract ambient noise in an area. Workers in noisy environments may favor this method over ear plugs.
- Noise reduction: In sound and video equipment, noise reduction is the process of removing noise from a signal. This is strictly for electronic noise or noise which has been detected and put into electronic form.
- Noise canceling: If both the noise and the signal are received by an electronic or digital medium, noise can be filtered from the signal electronically and retransmitted without the noise. Helicopter pilots rely on this technology to speak on the radio.

HEARING PROTECTION DEVICE

A hearing protection device, also known as a HPD, is an ear protection device worn in or over the ears while exposed to hazardous noise to help prevent noise-induced hearing loss. HPDs reduce (not eliminate) the level of the noise entering the ear. HPDs can also protect against other effects of noise exposure such as tinnitus and hyperacusis. There are many different types of HPDs available for use, including earmuffs, earplugs, electronic hearing protection devices, and semi-insert devices.

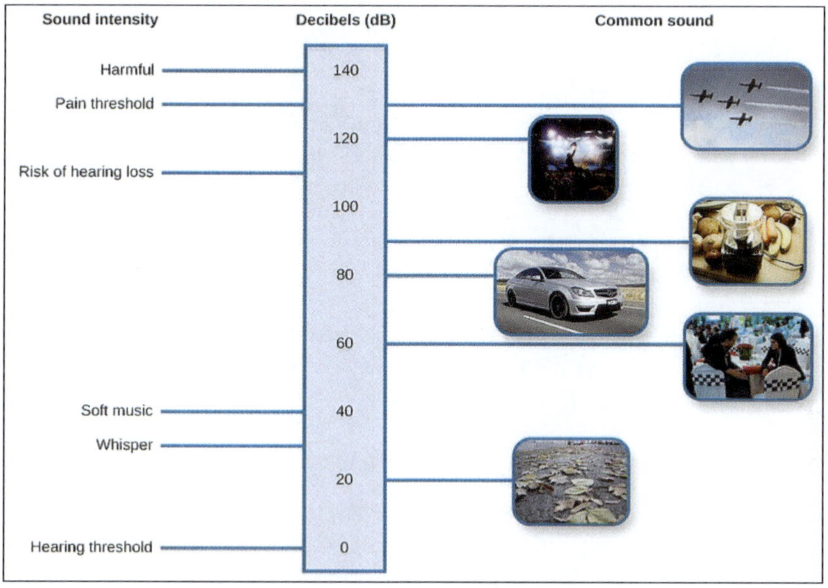

Sound levels of some daily activities.

Types

Earmuffs

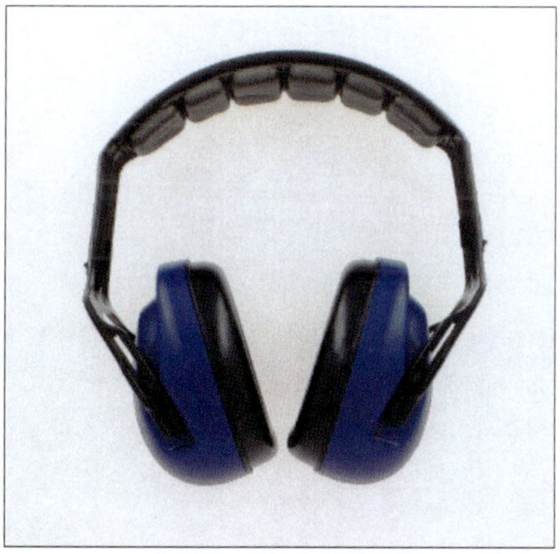

Earmuff hearing protection device.

Earmuff style hearing protection devices are designed to fit over the outer ear, or pinna. Earmuff HPDs typically consist of two ear cups and a head band. Ear cups are usually lined with a sound-absorbing material, such as foam. The cups should be fit so that the center of the ear canal aligns with the ear canal opening. The soft cushions seal around the pinna of the ears. The head band, centered at the top of the head, applies force/ pressure to seal the ear cups over the ears.

Earplugs

Different styles of earplugs are pictured. Left, pre-molded earplugs.
Center, formable earplugs. Right, roll-down foam earplugs.

Earplug style hearing protection devices are designed to fit in the ear canal. Earplugs come in a variety of different subtypes. The attenuation offered by these devices can be measured through hearing protection fit testing.

- Pre-molded earplugs have a preformed shape and a push-to-fit design.
- Formable earplugs are pliable and take the form of an individual's ear canal.
- Roll-down foam earplugs are one of the most commonly used earplugs, and are made from slow recovery foam which expands after it has been "rolled-down" and inserted in the ear canal, creating a tighter seal.
- Custom Earplugs are made individually for each user following ear impressions. Typically custom earplugs are purchased from an audiology clinic or hearing healthcare professional.

Electronic Hearing Protection Devices

Some HPDs reduce the sound reaching the eardrum through a combination of electronic and structural components. Electronic HPDs are available in both earmuff and custom earplug styles. Electronic microphones, circuitry, and receivers perform active noise reduction, also known as noise-cancelling, in which a signal that is 180-degrees out-of-phase of the noise is presented, which in theory cancels the noise.

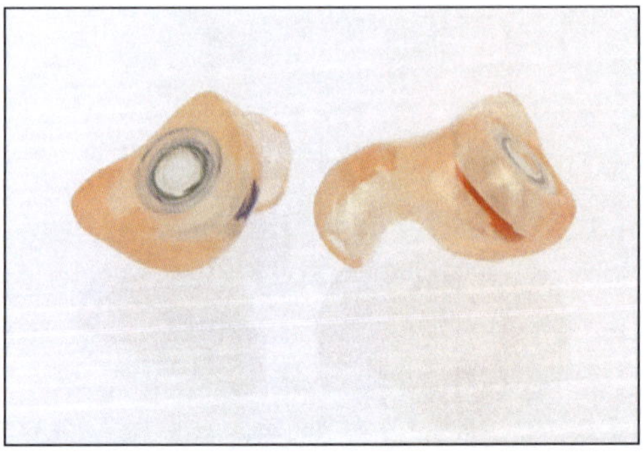

Custom earplug hearing protection devices.

Some electronic HPDs, known as Hearing Enhancement Protection Systems, provide hearing protection from high-level sounds while allowing transmission of other sounds like speech. Some also have the ability to amplify low-level sounds. This type may be beneficial for users who are in noisy environments, but still need access to lower level sounds. For example, soldiers who need to protect their hearing but also need to be able to identify enemy forces and communicate in noise, hunters who rely on detecting and localizing soft sounds of wildlife but still wish to protect their hearing from recreational firearm blasts, as well as users with pre-existing hearing loss who are in noisy environments may all benefit from Hearing Enhancement Protection Systems.

Electronic HPDs require the use of batteries and are typically more expensive than non-electronic types.

Semi-insert Devices (Canal Caps)

Canal caps are similar to earplugs in that they consists of soft tip that is inserted into the opening of the ear canal. Some styles are inserted slightly into the ear canal while others sit in place at the opening of the ear canal. In this case, the tips or caps are connected by a lightweight band which also serves to hold them in position.

Dual Hearing Protection

Dual hearing protection refers to the use of earplugs under ear muffs. This type of hearing protection is particularly recommended for workers in the Mining industry because they are exposed to extremely high noise levels, such as an 105 dBA TWA. Fortunately, there is an option of adding electronic features to dual hearing protectors. These features help with communication by making speech more clear, especially for those workers who already have hearing loss.

Hygiene and Care

In order to prevent irritation or infection of the ear, reusable HPDs should be cleaned on a regular basis. Before using any HPD, it should be inspected for damage or dirt to ensure that it is safe to use. Single-use, disposable earplugs are available in addition to reusable options. Earplugs intended for single-use should not be washed for reuse as this degrades the material and reduces effectiveness.

Most reusable earplugs can be cleaned using mild soap and warm water between uses and should be replaced every 2-4 weeks. Earmuff cups and cushions should be cleaned regularly with soap and water, and be replaced if they become cracked or otherwise compromised. Ear cushions can last from 3-8 months depending on use. Use of a clean, protective case to store your HPDs when not in use is recommended to prevent damage or contamination.

Any damage to a HPD can compromise its integrity, thus reducing its effectiveness. Damaged HPDs should not be used.

Recommended Use

Many countries require several interventions to control risks from exposures to loud noise in the workplace. In the US, the Occupational Safety and Health Administration requires hearing conservation programs which include the provision of hearing protection devices. It is also recommended by the U.S. National Institute for Occupational Safety and Health, audiologists and other hearing healthcare professionals when one works exposed to noise levels that exceed 85 dB. NIOSH and OSHA base their recommendations for use of hearing protection by a calculation called time-weighted average (TWA). A time-weighted average is the average noise level a worker is exposed to over a period of time. NIOSH recommends that OSHA use an 85 dBA time-weighted average during an 8 hour period as their exposure limit. An 85 dBA time-weighted average means that HPD use is recommended if an employee is exposed to an average noise level of 85 dBA or more during an 8-hour work day. NIOSH also uses a 3 dB exchange rate for time-weighted averages. A 3 dB exchange rate means that for every 3 dB increase in the average level of noise the recommended time being exposed to that level of noise is cut in half. For example, for a worker who is exposed to 88 dBA, it's recommended he/she only be exposed to that level of noise for 4 hours. These levels of noise may be encountered in both occupational and recreational settings. HPDs are recommended for use in settings where it is difficult to control the noise level, and the person exposed to the noise cannot be removed from the environment.

The amount of protection from noise can vary based on the physical fit of the device. Hearing protection devices with accurate placement (an airtight seal) and/or accurate insertion (deep into the ear canal) will provide the most attenuation of noise. There are many challenges to achieving the needed protection from the device, from barriers to adequate use, to issues related to comfort, convenience, lack of training, to beliefs and attitudes towards its use.

Noise Reduction Ratings

Hearing protection device manufacturers in the United States are required by the EPA to label HPDs with a noise reduction rating, or NRR. The NRR estimates how much noise is reduced by a hearing protection device, measured in decibels.

The NRR is measured by manufacturers using American National Standards Institute (ANSI) specified procedures in a laboratory environment. The NRR tends to overestimate the amount of protection the HPDs provide in real-world conditions. These differences are most likely attributed to incorrect insertion or poor earplug fit. Because the actual amount of attenuation is typically less than the labeled NRR, the U.S. Department of Labor/OSHA suggest "derating" or reducing the NRR when using it to evaluate total noise exposure when wearing the HPD. OSHA's training manual has uses a 7 dBA correction factor: NRR- 7dBA= estimated noise exposure.

In the late 90's the US National Institute for Occupational Health and Safety (NIOSH) recommended a different derating scheme for HPDs but now encourages the hearing protection fit-testing. Hearing protection fit-testing has been developed in order to determine the actual attenuation of the device as it is worn for an individual. These tests for checking attenuation values summarize the real-world attenuation in a personal attenuation rating (PAR). The PAR is unique to the HPD tested and the individual wearing the protection. When obtaining a PAR, it is recommended to test both ears individually as there may be asymmetry between ears. This asymmetry can be caused by anatomical differences between ears or improper fit of the HPD and could lead to a unilateral threshold shift. A PAR can be reported as C-weighted values, A-weighted values, or attenuation values for specific frequencies. The PAR can then be used to determine if the HPD is providing adequate attenuation so that an individual's noise exposure does not exceed the recommended limits set forth by regulatory agencies such as the Occupational Safety and Health Administration (OSHA).

NOISE BARRIER

The *sound tube* in Melbourne, Australia, designed to reduce roadway noise without detracting from the area's aesthetics.

A noise barrier (also called a soundwall, noise wall, sound berm, sound barrier, or acoustical barrier) is an exterior structure designed to protect inhabitants of sensitive land use areas from noise pollution. Noise barriers are the most effective method of mitigating roadway, railway, and industrial noise sources – other than cessation of the source activity or use of source controls.

In the case of surface transportation noise, other methods of reducing the source noise intensity include encouraging the use of hybrid and electric vehicles, improving automobile aerodynamics and tire design, and choosing low-noise paving material.

Extensive use of noise barriers began in the United States after noise regulations were introduced in the early 1970s.

Design

The acoustical science of noise barrier design is based upon treating an airway or railway as a line source. The theory is based upon blockage of sound ray travel toward a particular receptor; however, diffraction of sound must be addressed. Sound waves bend (downward) when they pass an edge, such as the apex of a noise barrier. Barriers that block line of sight of a highway or other source will therefore block more sound. Further complicating matters is the phenomenon of refraction, the bending of sound rays in the presence of an inhomogeneous atmosphere. Wind shear and thermocline produce such inhomogeneities. The sound sources modeled must include engine noise, tire noise, and aerodynamic noise, all of which vary by vehicle type and speed.

The noise barrier may be constructed on private land, on a public right-of-way, or on other public land. Because sound levels are measured using a logarithmic scale, a reduction of nine decibels is equivalent to elimination of approximately 86 percent of the unwanted sound power.

Noise barrier earth berm along California State Route 12, Sonoma County, California.

Materials

Several different materials may be used for sound barriers. These materials can include masonry, earthwork (such as earth berm), steel, concrete, wood, plastics, insulating wool, or composites. Walls that are made of absorptive material mitigate sound differently than hard surfaces. It is now also possible to make noise barriers with active materials such as solar photovoltaic panels to generate electricity while also reducing traffic noise.

A wall with porous surface material and sound-dampening content material can be absorptive where little or no noise is reflected back towards the source or elsewhere. Hard surfaces such as masonry or concrete are considered to be reflective where most of the noise is reflected back towards the noise source and beyond.

Noise barriers can be effective tools for noise pollution abatement, but certain locations and topographies are not suitable for use of noise barriers. Cost and aesthetics also play a role in the choice of noise barriers. In some cases, a roadway is surrounded by a noise abatement structure or dug into a tunnel using the cut-and-cover method.

Tradeoffs

This noise abatement wall in The Netherlands has a transparent section at the driver's eye-level to reduce the visual impact for road users.

Potential disadvantages of noise barriers include:

- Aesthetic impacts for motorists and neighbors, particularly if scenic vistas are blocked;
- Costs of design, construction, and maintenance;
- Necessity to design custom drainage that the barrier may interrupt.

Normally, the benefits of noise reduction far outweigh aesthetic impacts for residents protected from unwanted sound. These benefits include lessened sleep disturbance, improved ability to enjoy outdoor life, reduced speech interference, stress reduction, reduced risk of hearing impairment, and a reduction in the elevated blood pressure created by noise (that improves cardiovascular health).

Sound barrier walls vary in cost depending on the type and quality. Concrete is popular due to lower cost. Since they are reflective, they could potentially create noise for those across from the barrier. Absorptive barriers absorb, and thus abate the noise more effectively, but cost more as they are often custom built and are composed of a variety of materials that allow for absorption. With regard to berm construction costs, a major factor is the availability of excess soil in the immediate area which could be used for berm construction. If the soil is present, it is often cheaper to construct an earth berm noise barrier than to haul away the excess dirt, provided there is sufficient land area available for berm construction. Generally a four-to-one ratio of berm cross sectional width to height is required. Thus, for example, to build a 6-foot-high (1.8 m) berm, one needs an available width of 24 feet (7.3 m).

Effects on Pollution

Roadside noise barriers have been shown to reduce the near-road air pollution concentration levels. Within 15–50 m from the roadside, air pollution concentration levels at the lee side of the noise barriers may be reduced by up to 50% compared to open road values.

Noise barriers force the pollution plumes coming from the road to move up and over the barrier creating the effect of an elevated source and enhancing vertical dispersion of the plume. The deceleration and the deflection of the initial flow by the noise barrier force the plume to disperse horizontally. A highly turbulent shear zone characterized by slow velocities and a re-circulation cavity is created in the lee of the barrier which further enhances the dispersion; this mixes ambient air with the pollutants downwind behind the barrier.

MUZZLE SHROUD

A muzzle shroud, linear compensator or concussion reduction device is a sleeve (either circular or otherwise) that attaches to and extends beyond the muzzle of a firearm to redirect some of the noise and concussion (shock wave) from the muzzle blast forward and away from the shooter and others standing behind and to the sides.

An MG-42 medium machinegun with a shrouded muzzle and barrel.

A muzzle shroud is similar to a barrel shroud with the difference that it only surrounds the far end of the barrel known as a muzzle. It can be a muzzle extension or muzzle device. It could be considered a flash suppressor without circumferential openings.

A muzzle shroud differs from a sound suppressor in that its primary function is not to suppress the sound, but merely redirect it so to increase the comfort of the shooter, reducing the concussion to a comfortable level.

Linear compensators are also a form of muzzle shrouds. While most muzzle brakes redirect the gases to the side and rearward, most linear compensators redirect the gases forward. Linear compensators and suppressors do not have the disadvantages of a redirected muzzle blast; they actually reduce the blast by venting high pressure gas forward at reduced velocity.

A muzzle shroud can be used in conjunction with other muzzle devices (such as a muzzle brake, flash hider or a brake/hider hybrid device). This way the user can still benefit from some of the recoil reduction of a muzzle brake, but with less of the noise and concussion normally associated with such a device. However, depending on the design, the CRD will usually increase the recoil to some degree compared to using a muzzle brake only.

Motivation

Even wearing double hearing protection (earplugs and muffs), some shooters have experienced hearing loss from high volume or long term shooting, probably due to the shockwaves from muzzle blasts propagating though the sinuses and facial bone structures. The noise levels experienced are often increased when firing indoor or in other confined spaces. Sound suppressors are increasingly commonly used by shooters to reduce the concussion and thereby prevent hearing loss, but are not always suitable for either legal or technical reasons. Suppressors can either be difficult or illegal to obtain in some jurisdictions, while a concussion reduction device is normally not classified as such. Concussion reduction devices also typically don't have the problem of introducing the extra amount of mirage as a suppressor does, since suppressors work by absorbing heat from the burnt gun powder. The extra mirage can make accurate aiming difficult, especially on precision targets at long range.

A muzzle shroud can also increase comfort compared to a muzzle brake while shooting prone by preventing dust to kick up from the gound.

Amount of Noise Reduction

Sound waves are semi-directional. They consist of both Point- and Line-wave propagation. The projectile emanates mostly line-propagation properties after leaving the barrel. At the moment that the projectile is exiting the barrel it produces a mainly point-biased wave generation. (This is largely due to the shock waves traveling forward along the interior of the barrel, thus preventing any backward travel from the projectile into the barrel.) Sound waves travel in an ever-expanding bubble emanating from the muzzle when the propellant gases are traveling at supersonic speeds. This is called the "muzzle blast." Except in extremely low pressure loads, the exiting propellant gasses generate much higher sound pressure levels (dBa) than any other aspect of the weapon firing. (e.g., hammer, sear, bolt, supersonic flight signature, etc.) The supersonic "ballistic crack" is typically 136-138 dB, whereas "muzzle blast" may reach 168 dB. By encircling the muzzle, the shroud funnels and focuses the sound waves along the projectile flight axis, away from the shooter. This alone can partially protect the shooter from 12-18 dB of blast returning to the shooter's ears.

HUSH KIT

A hush kit is an aerodynamic device used to help reduce the noise produced by aircraft jet engines. These devices are typically installed on older turbojet and low-bypass turbofan engines, as they are much louder than later high-bypass turbofan engines.

A hush kit installed on a Swedish Air Force Sud Aviation Caravelle.

Hush kits are used because the engines on jet aircraft generate a large amount of noise, which significantly increases noise pollution near airports.

Design

The most common form of hush kit is a multi-lobe exhaust mixer. This device is fitted to the rear of the engine and mixes the jet core's exhaust gases with the surrounding air and a small amount of available bypass air. Modern high-bypass turbofan engines build on this principle by utilizing available bypass air to envelop the jet-core exhaust at the rear of the engine, reducing noise.

Most hush kits make further modifications to the exhaust, including acoustically treated tailpipes, revised inlet nacelles and guide vanes. They reduce the forward propagating, high-pitched noise caused by the small, high-speed fan.

This kind of high-pitched noise is much less of an issue on modern high-bypass turbofan engines as the significantly larger front fans they employ are designed to spin at much lower speeds than those found in older turbojet, and low-bypass turbofan, engines.

Use

Modern aircraft equipped with high-bypass turbofan engines are designed to comply with contemporary aviation noise abatement laws and ICAO regulations. Several older aircraft that are still in service (typically in a cargo capacity) have hush kits retrofitted so that they are able to conform with noise regulations needed to operate in many commercial airports. Some of the examples include:

- Boeing 707
- Boeing 727
- Boeing 737-200
- Douglas DC-8
- Douglas DC-9
- Tupolev Tu-154

Hush kits can also be found on small business jets and other aircraft that are too small to be fitted with bulky high-bypass turbofan engines. In many cases, these aircraft are manufactured with hush kits installed as a more economical way to meet noise restrictions than what would be needed to make expensive engine or design changes. The Gulfstream II and Gulfstream III are examples of these aircraft.

Impact

Hush kits can adversely affect the range and performance of the aircraft they are fitted to because of the extra weight. It also reduces engine performance and aerodynamic efficiency. For example, the hush kit fitted to the Gulfstream II adds 106 kilograms (234 lb) to the total airplane weight of 65,500 lb (29,700 kg), causing around a 1.6% reduction in aircraft range.

Meanwhile, the hush kit for the Gulfstream G-III jets weighs about 170 kg (370 lb) and cause a calculated 2% drop in range, although the reduction in range has not been noticed during actual flights, which are rarely made to the very limit of the aircraft's range anyway.

On a larger aircraft, like Federal Express's Boeing 727s, the hush kits add 410 kg (900 lb) of extra weight (total airplane weight up to 86,000 kg (190,000 lb)) and this results in an overall 0.5% increase in fuel burn for short trips (but no measurable increase for long flights).

MASS LOADED VINYL

Mass Loaded Vinyl or MVL is a sound barrier blocking material known as a "Limp Mass Barrier."

The material is comprised of two principle ingredients: vinyl — to give the MLV flexibility, "the limpness" and, a naturally occurring, high mass element. Often Barium Sulfate is used due to its unique properties – it is non-toxic and has a high relative density. This latter attribute of high density is what makes Mass Loaded Vinyl so effective in blocking sound.

Difference between Mass Loaded Vinyl and Acoustic Sound Batts

We first need to understand what makes a good absorber or barrier. In general there are only two characteristics that make an effective sound absorber or sound barrier – "fuzzy stuff" (mineral fibre or fiberglass) or mass. Most if not all absorbers and barriers have one or both of these characteristics in their design. How these materials are employed in the design of the absorber/barrier or, how and where the absorber/barriers are placed in an environment is the wizardry of acoustic design.

Absorbing sound relies heavily on the knowledge and manipulation of a sounds wavelength. For example, mid to high-frequency sound — say above 500 Hz — have wavelengths greater than 68 cm in length (˜ 27 inches). To absorb this wavelength with fuzzy stuff would require a minimum depth of absorber of one-quarter of the waves length. In this example, 17 cm (˜ 6 inches) of depth would be needed to attenuate frequencies at and above 500 Hz.

Fuzzy stuff absorbers function by virtue of the fuzzy materials porosity (the percentage of pores or crevices, to the total volume of the mass). These crevices capture the relatively short wavelengths of mid to high-frequency sound and convert the sounds energy to heat via friction.

Blocking sound on the other hand, requires that the material to be used be the opposite in composition to an absorber. The material needs to be dense. Density is derived from mass. Highly dense or massive materials typically do not have a porous structure to allow the sound wave to enter (and pass through) them. Consequently, any material that has a high mass could be used to block a sounds path. Concrete, gypsum board, MDF, hardboard and concrete masonry units are all popular examples of high mass barriers. The problem with these common building materials is that they are very stiff. They have great mass but they don't get the overall sound reductions you would expect from them because they are not limp and flexible like the mass loaded vinyl (MLV) Noise Blok Pro. Again it is the application of the mass (the what, where and how the mass is used) that will make an effective sound barrier.

Because MLV has a very small cross-section, it's flexible AND it's massive.

None of the earlier examples of high mass barriers can be wrapped around or molded to the shape of something — MLV can. This unique combination of small dimension, mass and flexibility makes Mass Loaded Vinyl a one-of-a-kind material in the world of sound barriers.

For example, if you need to sound isolate a horizontal run of a grey-water pipe in a homes basement; the typical method in the past would have been to place a little more fuzzy stuff in the joist cavity where the pipe is (say 3-inch Roxul Safe n' Sound insulation). Then put up the normal 1/2-inch gypsum board ceiling.

If we examine the weight per square foot of this single-layer construction, we would find that there is a total of 25.6 oz./sq. ft. for the 1/2 inch gypsum board plus approximately 10 oz./sq. ft. for the 3-inch Roxul. A total of approximately 2.2 lbs./sq. ft.

Installing more layers of gypsum via means of Damping Compound or isolation clips is unfortunately not always doable due to the floor to ceiling height in many basements. Therefore, we find that this is the typical total isolation for the grey-water pipe noise. Soundproofing walls and soundproofing ceilings in these restricted environments is often far easier with Mass Loaded Vinyl. This is where Mass Loaded Vinyl is irreplaceable.

Going back to the earlier example of the grey-water pipe. Before closing the ceiling, if we wrap the grey-water pipe in .14 mm (˜ 9/64-inch) Mass Loaded Vinyl at 2 lbs./sq. ft., we have effectively doubled the mass per running foot of the ceiling around and below the pipe.

Double the mass = approximately double the isolation. Try wrapping a pipe in concrete or hardboard or MDF.

Even if it where doable — think of the time in labour that would be required? Compared to cutting lengths of Mass Loaded Vinyl into strips and then cable tying the strips around the pipe. It's a bit of a no-brainer.

Installing Mass Loaded Vinyl in walls, ceilings and beneath floors is a space saver and is stunningly effective regarding mass per cross-section dimension. Where to use Mass Loaded Vinyl?

- Soundproofing walls;
- Soundproofing ceilings;
- Soundproofing floors;
- Soundproofing pipes;
- Soundproofing HVAC ducts;
- Soundproofing office partition walls;
- Soundproofing computer server rooms;
- Soundproofing mechanical rooms.

Acoustic Panelling

Acoustic panels are generally between 40 and 200 mm thick and are made from a sandwich of coated steel sheets, with a mineral fibre infill. On one side of the panel the metal sheet is perforated to allow sound to pass inside where it is absorbed by the in-fill material. Acoustic panels can be designed to offer very high attenuation to noise transmission that is better than an equivalent solid structure of the same weight. Also, their absorptive properties are used to control reverberation in confined spaces.

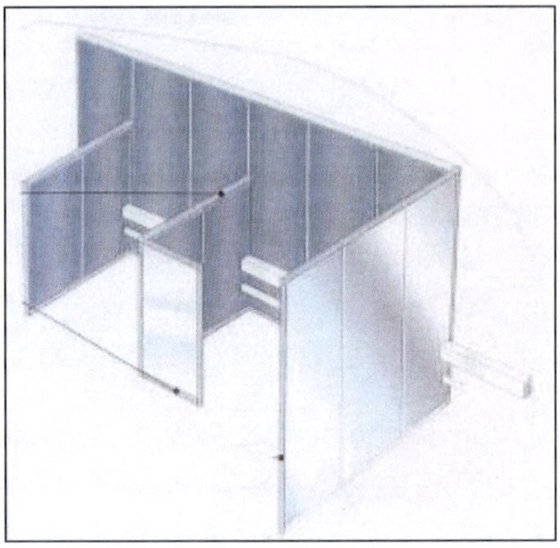

There are many types of acoustic panel, but most are designed to absorb sound incident from one side.

Application

Acoustic Panels are the building blocks of many acoustic products. Used alone they can replace conventional walls or partitions, or used to line plant rooms within existing structures. Attenuation to sound transmission can be up to about 40 dB(A). Absorption can be over 90% at middle and high frequencies.

Acoustic Lagging

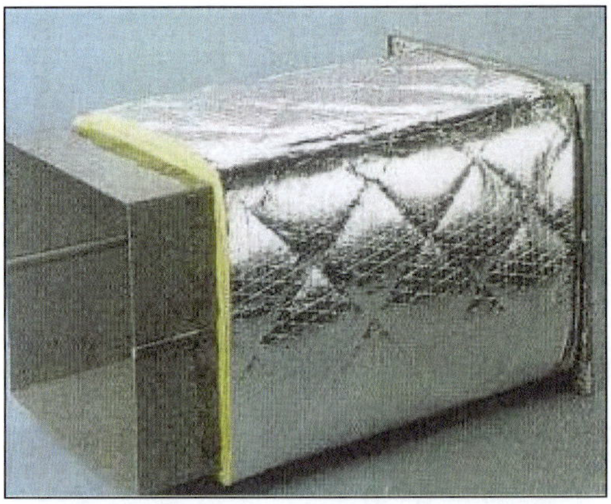

A flexible material wrapped around noise sources in the same way as thermal lagging.

Acoustic lagging is generally made of a laminated quilt comprising a tough outer skin, mineral fibre quilting and a heavy (often lead) internal layer providing mass. Such lagging is flexible and can be fitted around complex shapes and is ideal for sealing holes and gaps around solid acoustic elements. A similar product can be hung as acoustic curtain around plant that is moved frequently of in very confined spaces.

Application

Lagging to pipes and ducts. Sealing around most acoustic installations. Acoustic curtain for pumps, fans, hand tools etc. Performance is 10 to 15 dB(A).

Vibration Damping

Metal sheeting, has a tendency to 'ring' when vibrated by heavy machinery. Large unsupported areas of metal can be very efficient at converting vibrational energy into audible frequencies and radiating noise, rather like a loud speaker cone. Casings and guards can be damped using various damping or 'antidrumming' compounds.

Damping materials applied as an adhesive, flexible 'putty' or sheeting to vibrating metal sheeting.

Application

Casings and guards to most machinery, e.g., conveyors, pumps, rotating parts.

Performance

Performance varies, but 10 dB(A) is achievable. Damping of machinery casing is most effective on large areas of thin gauge metal.

Impact Deadening

Deadening materials can be useful on metal floors, chutes or containers, but there are practical problems with wear and tear. For metal floors blended resin compounds have been developed that damp noise generated by trolley wheels and foot falls whilst offering adequate grip, strength and weather proofing. Metal chutes and lorry holds can be surfaced with abrasion-resistant rubber to reduce noise generated by material impacts.

Resilient materials can also be used to help reduce noise generated by inherently noisy material, e.g., barrels, scrap metal and pipes.

Noise generated by impacts on metal surfaces such as loading material into dump trucks, can be reduced by applying a resilient surface treatment.

Application

Metal floors or wall coverings in warehouses or industrial stores. Lorries, metal chutes, hoppers. Impact noise can be reduced by 10 dB(A) or more.

Attenuators

Attenuators can be part of a ducted air flow system or can be inserted in a wall, partition or enclosure opening. Several splitters divide the air flow. Splitters comprise perforated metal or rigid mesh enclosing an acoustic in-fill which is generally some form of mineral fibre. Problems include degradation of the infill material and pressure loss (energy consumption) due to restricting the air flow. Excessive air flow speed can regenerate noise. These problems do not prevent attenuators being very widely used in industry.

Attenuators are designed to allow air to flow between a series of splitters which absorb noise. They come in all shapes and sizes.

Application

Standard attenuators : Heating, ventilation and air conditioning systems, fans. Specialist attenuators: As used on motor vehicles, modified materials allow high temperature and corrosive gas discharges to be silenced.

Performance is a function of attenuator length and the ratio of the cross-sectional areas of the splitters and the whole attenuator. Up to 45 dB(A).

Steam and Air Diffusers

Steam traps and compressed air valves, used in petrochemical, food and other process plants to release excessive pressure, generate noise because of the high pressure and velocity of the discharge. The discharge flow can be passed through a diffuser of enlarged diameter filled with a flow resistant material such as stainless steel wool to slow down the gas speed, thus reducing noise generation.

Metal fittings attached to the outlets of high pressure steam or air discharge.

Application

Steam and high pressure air system discharges. Performance is best at mid to high frequencies and typically gives up to about 15 dB(A).

Vibration Isolation Mounts

Stiff springs or low compliance rubber/neoprene pads are housed in metal retaining fittings forming a mount or hanger to which the plant is bolted. The mount is designed to be deflected under static load and to absorb the predominant frequencies at which the plant oscillates. In practice this means selecting a mount whose natural frequency is well below the driving frequencies in the plant. Poor installation can lead to flanking paths, poor performance or amplification.

Spring or rubber mounts located between vibrating plant and their supporting structure reduce the transfer of vibrational energy into the structure where it may transferred and re-radiate as unwanted noise.

Application

Vibration isolation mounts can be fitted to most mechanical equipment, including pumps, fans, generators, compressors, and also to pipes, and ductwork.

Vibration isolation can be almost complete for correctly selected mounts. Installation is harder for large heavy plant and performance can be compromised at lower frequencies. Noise reduction performance will depend on the surrounding structure and its noise transfer and radiation characteristics.

Inertia Bases

Mounting equipment on spring or rubber isolators makes the system inherently unstable and can compromise the safe operation of the plant. These problems can usually be overcome with inertia bases which provide the following benefits; lower centre of gravity, reduced movement during start up, and even spreading of load.

Where plant is permitted a degree of freedom to move on such systems, flexible connectors may be required in pipework or other services to prevent fatigue and vibration transfer via flanking paths.

A massive, usually concrete, block onto which the plant is mounted via anti-vibration mounts or pads.

Application

Mainly applicable to large heavy machinery in combination with vibration isolation mounts or sometimes rubber/neoprene pads for extra stability.

NOISE MANAGEMENT PRACTICES

For some operational facilities there are effective ways of reducing noise simply by being aware of its presence as an issue for the site, and by adopting appropriate procedures when carrying out every day activities. Such procedures can be collectively called 'noise management' and can be particularly important where substantial noise control has been incorporated in a plant design. Noise management may be part of an overall Environmental Management System, in which case it may be subject to routine auditing and reporting requirements.

Routine Maintenance of Plant

Noise is usually generated in mechanical plant by the interaction of moving, and often rotating, parts. Over time these parts tend to wear, stretch or become distorted in some way and the levels of noise generated tend to increase. Common examples include bearings, gear chains, rollers and engines. Routine maintenance or servicing of such equipment can have a significant effect on noise output.

Noise control equipment can also degrade with time and require servicing or replacement. Engine silencers are a prime example. Acoustic panelling can become clogged, waterlogged and distorted and hence will become less effective.

Good Operational Site Practices

There are a number of common sense procedures that can help to reduce noise emissions. Although these tend to be specific to operations at particular facility some common examples are listed as follows:

- Closing doors and windows in noisy buildings and acoustic enclosures;
- Ensuring that generator or vehicle engine hatches are kept closed;
- Locating mobile plant away from noise sensitive receivers where possible;
- Avoiding dropping metallic materials from a height;
- Switching off plant when not in use;
- Stock piling materials (e.g., containers) so as to provide acoustic screening between noise sources and receivers; and
- Arranging delivery or on site vehicle routes away from sensitive receivers.

Although the noise reduction benefits of these practices can be difficult to quantify, they should form a routine part of best practice to reduce overall noise emissions.

Restricting Operating Hours

The sensitivity of neighbouring areas to noise impacts will vary with the time of day and on different days of the week/weekend. Restricting the operating hours of noisy activities can be an extremely effective way of mitigating community noise impacts and is often used, to great effect, in planning conditions for new facilities. Restricting operating hours can of course reduce productivity and create operational difficulties, but it need not necessarily require a complete cessation of all activity on the site. In some cases it will be possible to schedule noisy operations to the less sensitive daytime, weekday periods in order to keep noise emissions to a minimum at night.

Community Liaison

Some plants require occasional unavoidably noisy procedures to be followed such as ‘blowing down’ pressurised systems and testing emergency generators. There is evidence that the impact of noise from such activities can be reduced simply by warning the community beforehand. Often, the fact that a member of the public has been informed, and is aware that the operator is being as considerate as possible will reduce the level of disturbance caused. For one thing people will tend to schedule particularly noise-sensitive activities away from such periods. Of course the operator should try to schedule the noisy operations sympathetically too. An example of this is the testing of emergency generators in commercial areas where there are no residential buildings, on Sundays. More generally, keeping the community aware of noise control activities and plans will offer long term benefits to all parties, and should be encouraged.

Complaints due to noise should be treated in a constructive manner. They can often provide the operator with guidance on the main source of concern and the best approach to mitigating the problem. A general procedure for handling a noise complaint is as follows:

- The complaint is logged.
- The facility under consideration is visited immediately to inquire: If the operator knows what the source of the noise could be; and if the noise could be due to an unusual activity.
- The source of excessive noise is investigate by inspection to see if there is an obvious remedy, if so it is implemented and the complainant is updated.
- If no remedy is implemented, and excessive noise is suspected, noise monitoring is carried out to help identify the source and the extent of any noise impact.
- If monitoring identifies a noise impact remedial work is planned and implemented, the complainant is advised of progress.
- Repeat monitoring is carried out to establish whether the remedial work has been successful.
- The complainant is advised and the logged complaint is signed-off as having been properly addressed.

It is important that throughout this process the complainant is kept updated of progress, particularly if the investigations take some time.

Planning Noise Control

The ideal time to consider noise control is at the initial planning stage of a new facility. Similar opportunities may arise during the lifetime of a facility when planning an extension or when old plant is being replaced. At this stage potential noise problems can be 'designed out' rather than being left to be addressed later through a bolt on, end of pipe solution. This approach can not only produce more effective noise control but tends to be more cost effective too.

In some cases there will be planning restrictions governing what can be done on the site and these can limit the options for noise control, but in general consideration should be given the following general principals:

- Use of inherently quieter processes;
- Selection of inherently quiet plant or 'low noise options';

- Site layout to maximise natural screening, screening by buildings, and separation distances;
- Orientation of directional noise sources away from sensitive receivers; and
- Noise barriers and bunding.

In the particular case of landfill sites the design of the filling sequence can influence the extent of noise screening by filled material, and hence can be used to maximise screening for particular noise-sensitive areas.

Tree planting may provide effective mitigation of visual impacts but rarely reduces noise. For trees to have significant attenuation benefits they will need to be planted very close together and to cover a substantial propagation distance, for example a substantial mature woodland.

Costs, Benefits and Implications of Noise Control

In addition to the acoustic benefit achieved, long-term costs in both monetary and environmental terms are considered. It is not appropriate to attempt a cost benefits analysis of different noise control techniques because in most remedial situations the choice of noise control options will be limited by practical operational factors and there may be only one feasible solution.

Cost Benefits

The methods of noise control described in this report have generally been in widespread use for many years (with the exception of active noise control). It is because acoustic hardware provides tangible noise benefits for modest costs that it used so commonly. The installation of a suitable piece of acoustic hardware typically provides several decibels of attenuation and can achieve up to about 50 dB, although this is rare in practice. Of course, the reduction in the overall noise level from a facility tends to be less because there are usually many sources of noise that add up at the receiver point, all of which may potentially need acoustic treatment. Nonetheless, in environmental terms the noise benefits achieved represent significant improvements in the noise climate experiences by local communities. Such reductions are commonly achievable through tried and tested techniques. There are also benefits to the workforce who will experience a more pleasant work environment.

Costs

The cost of acoustic hardware varies hugely depending on the application. A simple fan silencer may cost a few hundred pounds, whereas it is not uncommon for power stations to incur noise control costs measured in millions of pounds. Compared to initial construction costs and ongoing operational costs, the costs of noise control are usually affordable, except to businesses operating on very low margins where only the cheaper methods can be used.

Noise control through good planning generally has minimal costs and potentially substantial noise reduction benefits. Ongoing noise management has minimal costs but generally the benefits are smaller. Figure summarises, in general terms, the relative cost benefits of some of the noise control methods that have been discussed in this report.

Noise management may be cost effective, but will rarely offer a complete solution to a serious noise problem. Low cost acoustic hardware, represented in the central part of the figure, may be affordable and may offer sufficient noise reduction in many cases, but in severe cases where acoustic enclosure and silencers are required costs will be higher. Active noise control is a relatively expensive method, but can provide a solution where other techniques cannot, ie for severe low frequency noise problems. This figure illustrates once again that planning for noise control will invariably offer the most cost-effective solution.

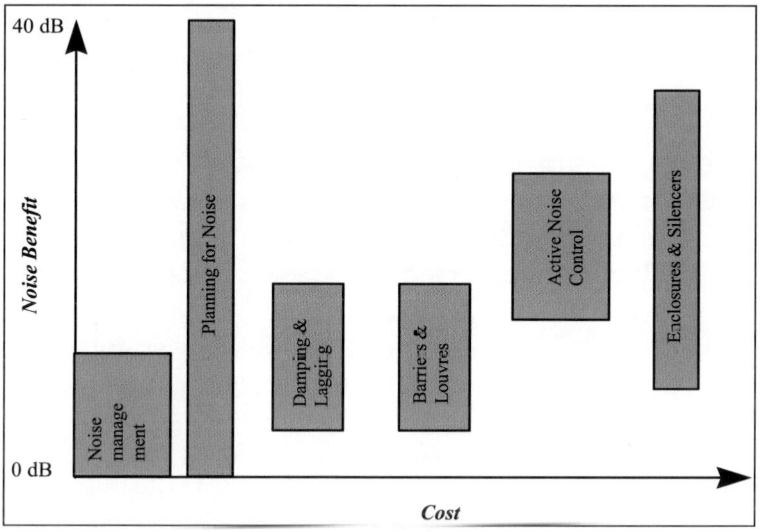

Cost Benefits of Noise Control.

General Implications

For an operational facility there will usually be restrictions as to which choice of noise control method to pursue because of the practicalities of retrofitting equipment and the long term implications of having it in place. There may be a lack of space, insufficient access for maintenance, safety implications or any number of difficulties to contend with.

Noise control equipment may also interfere with the correct operation of the plant, for example by placing it under increased load. Most noise control equipment works by blocking the path by which noise is emitted. This will tend to also block airflow and cooling which could increase wear in the plant and have long term maintenance and cost implications. These problems are not incurred by active noise control, although some maintenance will be required and speakers do wear out.

Noise management practices will rarely conflict with the safe operation of the plant as they are aimed at ensuring operatives use noise control features as they were designed to be used.

When planning a site layout with noise control in mind there will be numerous other considerations that will require careful consideration in achieving an optimal site and plant layout. However, in the past noise control has tended to be left until last and it should be considered along with all other factors in a balanced way during the conceptual and detailed design process.

Environmental Implications

Energy Consumption

Passive equipment added to a plant has the capacity to increase energy consumption by adding mass and restricting the free operation of the plant. Attenuators fitted to fans impede air flow and hence put additional load on the fan. This increased load, quantified as 'pressure drop' can be minimised by using a silencer with more open splitters but of greater length if there is space. Sometimes performance is increased by using two silencers in series. However, inevitably some extra load is created and more energy is required to drive the air flow. This effect is more severe when adding attenuators to high velocity air flows, such as in the inlet and discharge stacks to gas turbines in power stations. In high velocity air flows attenuators produce higher pressure drops which can have an appreciable effect on the efficiency of the gas turbine and hence the whole power generation process. For example, a stack silencer producing a pressure drop of 30 millibars can reduce overall power generating efficiency by 2%. Consequently acoustic engineers work to tight guidelines when designing and selecting such attenuators.

Raw Material Usage

The manufacture, servicing, replacement and decommissioning of acoustic hardware all require natural resources. There are clear advantages in choosing an inherently quiet process or low noise plant option rather than bolting on noise control equipment. Where acoustic hardware creates additional mechanical or heat load the life of the plant may potentially be shortened which will also have resource implications. These will not be major concerns for smaller installations, but they may be more important in sectors such as the power industry where more substantial resources are involved.

Acoustic hardware generally has a lifetime of many years and is largely maintenance free. As such it generates very little waste when operational. Acoustic panels can degenerate under adverse conditions, for example if they became very wet the infill material may need replacing from time to time. There has been concern that very fine mineral fibres can be hazardous, but safer products have been developed.

References

- Hansen, colin; snyder, scott; qiu, xiaojun; brooks, laura; moreau, danielle (2012), active control of noise and vibration, second edition, crc press, p. 3, isbn 9781482234008
- Principles-and-techniques-of-noise-control, essay: yourarticlelibrary.com, Retrieved 21 May, 2019
- Wagner, kate (8 december 2016). "building the wall: highway sound barriers and the evolution of noise". 99 percent invisible. Retrieved 21 march 2017
- Moylan, william (2006). Understanding and crafting the mix: the art of recording. Focal press. P. 26. Isbn 0-240-80755-3
- What-is-soundproofing: soundblock.com.au, Retrieved 8 January, 2019
- Schröder, ernst f. "the story of high com". Archived from the original on 2016-04-16. Retrieved 2016-04-16
- Mass-loaded-vinyl, soundproofing-walls-ceilings, products: acoustiguard.com, Retrieved 13 May, 2019
- Rawool, vishakha waman (2011). "chapter 6: hearing protection and enhancement devices". Hearing conservation: in occupational, recreational, educational, and home settings. Thieme. Pp. 136–173. Isbn 978-1604062571

6

Sound Level Measurement

Various instruments are used for the measurement of sound level. The most common of them are microphones, sound level meters, frequency analyzers, noise dosimeters, personal sound level meters, calibrators, etc. All these diverse methods and instruments of sound level measurement have been carefully discussed in this chapter.

SOUND MEASURING INSTRUMENTS

Many types of measuring systems can be used for the measurement of sound depending on the purpose of the study, the characteristics of sound and the extent of information that is desired about the sound. The various elements in a measuring system are:

- The transducer; that is, the microphone;
- The electronic amplifier and calibrated attenuator for gain control;
- The frequency weighting or analyzing possibilities;
- The data storage facilities;
- The display.

Not all elements are used in every measuring system. The microphone can, for instance, be connected to a sound level meter or directly to a magnetic tape recorder for data storage and future measurement or reference. An example of the components of the sound level meter is shown in figure.

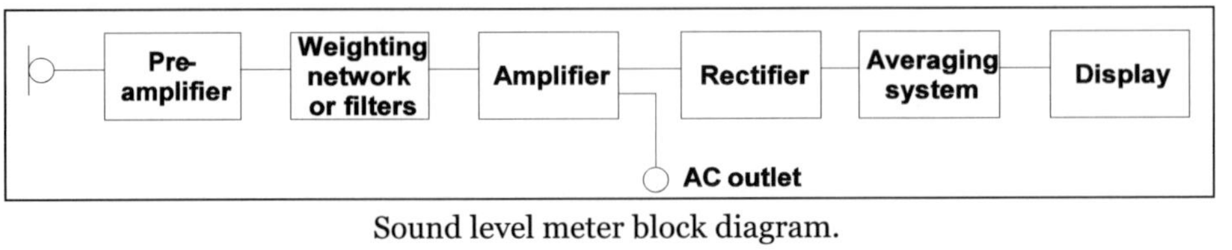

Sound level meter block diagram.

The two main characteristics are:

1. The frequency response: That is, the deviation between the measured value and the true value as a function of the frequency. As the ear is capable of hearing

sounds between 20 Hz and 20 kHz, the frequency response of the sound level meter should be good, with variations smaller than 1 dB, over that range.

2. The dynamic range: That is, the range in dB over which the measured value is proportional to the true value, at a given frequency (usually 1000 Hz). This range is limited at low levels by the electrical background noise of the instrument and at high levels by the signal distortion caused by overloading the microphone or amplifiers.

Microphones

Different Types

The microphone is the interface between the acoustic field and the measuring system. It responds to sound pressure and transforms it into an electric signal which can be interpreted by the measuring instrument (e.g. the sound level meter). The best instrument cannot give a result better than the output from the microphone. Therefore, its selection and use must be carefully carried out to avoid errors. When selecting a microphone, its characteristics must be known so that its technical performance (e.g. frequency response, dynamic range, directivity, stability), in terms of accuracy and precision, meets the requirements of the measurement in question, taking into account the expected conditions of use (e.g. ambient temperature, humidity, wind, pollution).

The microphone can be of the following types: piezoelectric, condenser, electret or dynamic. In a piezoelectric microphone, the membrane is attached to a piezoelectric crystal which generates an electric current when submitted to mechanical tension. The vibrations in the air, resulting from the sound waves, are picked up by the microphone membrane and the resulting pressure on the piezoelectric crystal transforms the vibration into an electric signal. These microphones are stable, mechanically robust and not appreciably influenced by ambient climatic conditions. They are often used in sound survey meters.

In a condenser microphone, the microphone membrane is built parallel to a fixed plate and forms with it a condenser. A potential differential is applied between the two plates using a D.C. voltage supply (the polarisation voltage). The movements, which the sound waves provoke in the membrane, give origin to variations in the electrical capacitance and therefore in a small electric current. These microphones are more accurate than the other types and are mostly used in precision sound level meters. However, they are more prone to being affected by dirt and moisture.

A variation on the condenser microphone which is currently very popular is the electret. In this case the potential difference is provided by a permanent electrostatic charge on the condenser plates and no external polarising voltage. This type of microphone is less sensitive to dirt and moisture than the condenser microphone with a polarisation voltage.

The last type is a microphone where the membrane is connected to a coil, centred in a magnetic field, and whose movements, triggered by the mechanical fluctuations of the membrane, give origin to a potential differential in the poles of the coil. The dynamic microphone is more mechanically resistant but its poor frequency response severely limits its use in the field of acoustics.

Sensitivity of a Microphone

The sensitivity of a microphone is defined as the amplitude (in mV) of the output signal for an incident sound pressure of amplitude 1 Pa (94 dB) at 1000 Hz. It can also be expressed in decibels by the following expression:

$$\text{Sensitivity} = 20\log_{10}\frac{Vp_0}{V_0 p}\,\text{dB re 1 V/ Pa}$$

Thus, a microphone giving an output signal V of 10 mV for a pressure signal p of 94 dB has a sensitivity of 10 mV/Pa or -40 dB. Here p_o = 1Pa and V_o = 1 volt.

Frequency Response

Good quality piezoelectric or condenser microphones have usually flat frequency response characteristics from 2 Hz to an upper limit which depends on their size. This limit is about 2 kHz for a 1” diameter microphone, 4 kHz for a 1/2” and 8 kHz for a 1/4” microphone. Below this limit, the frequency response is independent of the orientation of the microphone with respect to the noise source, and therefore the microphone can be held in any orientation. Above this limit, the frequency response will depend upon the direction of the sound wave on the microphone membrane.

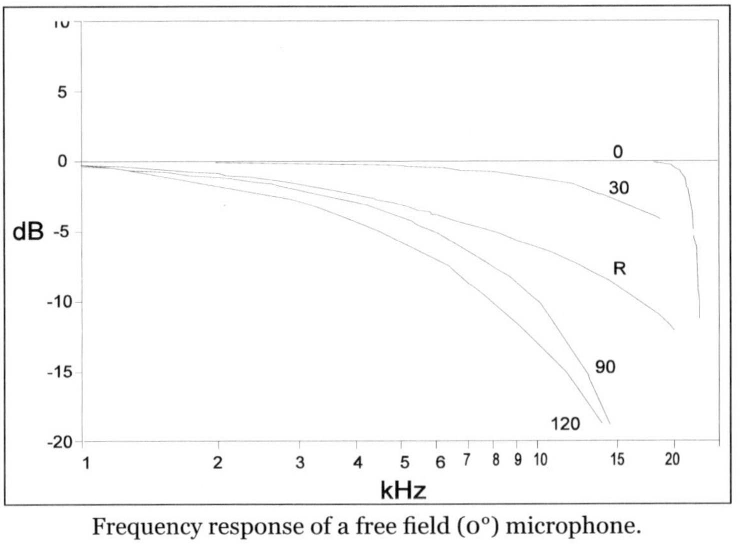

Frequency response of a free field (0°) microphone.

Some microphones have been designed in order for the response characteristics to be flat when the sound direction of propagation is perpendicular to the membrane. These microphones are called free field microphones and should be oriented toward the most significant sound source. Figure illustrates the frequency response characteristics of this type of microphone.

The numbers on the curves represent the angle of incidence (in degrees) of the incoming sound wave with respect to the normal to the membrane. The quantity, "R", represents the response to a diffuse sound field (sound incident equally from all possible directions).

Other microphones have been designed for the response characteristics to be flat when the sound comes in all directions at the same time as in a diffuse field. They are called diffuse field microphones. Their frequency response characteristic is very near the response characteristic under an incidence of 70° and these microphones should therefore be oriented at 70° toward the predominant sound source.

Figure illustrates the frequency response characteristics of this type of microphone.

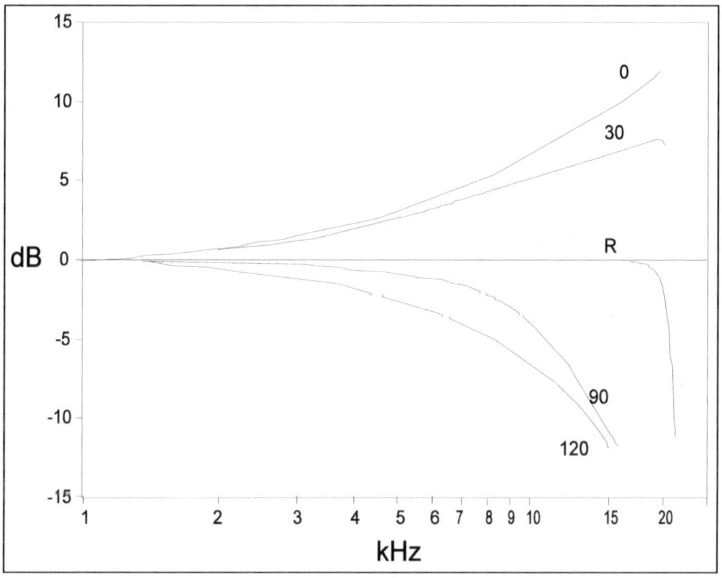

Frequency response of a diffuse field (R) microphone.

Dynamic Range

The output of a microphone is limited on the one hand by the internal noise of the transducer and on the other hand by the distortion resulting from high noisc levels. In addition, the instrument to which the output signal of the microphone is fed will saturate if the signal is too high and will also give a false result (that is, its background noise level) if the signal is too low. Therefore, high sensitivity microphones are needed to measure very low noise levels (lower than 30 dB), and low sensitivity ones have to be used for high noise levels such as for impact noise (above 130 dB). The dynamic range of typical good quality microphones is thus between 100 and 120 dB.

Selection and use of a Microphone

The selection of the microphone is based on:

- The levels to be measured,
- The frequencies to be measured - low or high,
- The type of acoustic field - free or diffuse,
- The purpose and the type of measurement - overall level or frequency analysis.

As stated, the measurement of low noise levels requires high sensitivity microphones and for high levels, low sensitivity ones are needed. The problem arises outside the range 50 to 120 dB usually: the characteristics indicated by the manufacturer for both the microphone and the indicating instrument should be checked.

If the noise is predominantly at frequencies below 1 kHz and overall levels are to be determined, any type of microphone may be used. On the contrary, if the noise is suspected or known to include a high frequency content, or that a frequency analysis is going to be made, the frequency characteristics of the microphone must be checked. As stated, the smaller the physical dimensions of the microphone, the wider the frequency range and the lesser the effects of directivity since they occur at higher frequencies. Microphones of small diameter would then be preferable. However, they are more fragile and, with the exception of certain special ones, less sensitive.

The user must then choose between free field or diffuse field microphones. When it is necessary to measure the ambient noise level at a given point regardless of the localisation of the sources or in presence of a diffuse field, a diffuse field microphone must be used: this is generally the case in occupational hygiene for the evaluation of exposure to noise.

On the contrary, for control purposes, the aim is usually to characterize the noise emitted by a particular machine. The machine should ideally be placed in a free field environment or at least in a very absorbing room and a free field microphone should be selected. If this is not possible, there exist ISO standards to assist with making the measurements "on-site" (ISO 3740, ISO11200).

Where possible, a diffuse field microphone should be directed at about 70° from the direction of the "predominant" noise source, so that the frequency responses for direct and reflected waves are the same. It should always be remembered that this concerns only high frequencies; therefore the "predominant" source is, in this regard, the one emitting the most at these frequencies, regardless of what is emitted at frequencies below 1 kHz.

Similarly, a free field 0° microphone must always be pointed toward the "predominant" source.

Sound Level Meters

The electrical signal from the transducer is fed to the pre-amplifier of the sound level meter and, if needed, a weighted filter over a specified range of frequencies. Further amplification prepares the signal either for output to other instruments such as a tape recorder or for rectification and direct reading on the meter.

The rectifier gives the RMS value of the signal. The RMS signal is then exponentially averaged using a time constant of 0.1 s ("FAST") or 1 s ("SLOW") and the result is displayed digitally or on an analog meter.

In some cases, the sound level meter does not include a logarithmic converter. The scale on the indicating device is then exponential so that the linear signal may be read in dB. In this case, the dynamic range of the display is usually restricted to 10 to 16 dB and the precision of the reading is rather poor. In the case of intermittent noise, the user must constantly adjust the amplifier to adapt the output signal to the dynamic range of the display.

When a log converter is used, the display scale is linear in dB and its dynamic range is usually much greater. This type of display has the advantage of providing the same precision at any level and permitting a much better appreciation of the range of fluctuations of the noise to be measured. In this regard, digital displays are less useful.

The specifications of sound level meters are given in IEC 60651 for four types 0, 1, 2, 3 differing by the measurement precision. The measurement precision is reduced as the type number increases, affecting manufacturing costs significantly. The IEC 60651 standard specifies the following characteristics:

- Directional characteristics;
- Frequency weighting characteristics;
- Time weighting, detector and indicator characteristics;
- Sensitivity to various environments.

The type 0 sound level meter is intended as a laboratory reference standard. Type 1 is intended especially for laboratory use, and for field use where the acoustical environment has to be closely specified and controlled. The type 2 sound level meter is suitable for general field applications. Type 3 is intended primarily for field noise survey applications. The frequency response for all types is defined from 10 Hz to 20000 Hz with a higher accuracy at frequencies from 100 Hz to 8000 Hz.

Type 2 and type 3 sound level meters usually include only the A-weighting network and the fast and slow response. Models with AC outlets should be chosen as they make it possible to record the noise on a magnetic tape recorder for further analysis. They are usually equipped with a diffuse field piezoelectric or electret microphone.

Type 0 and 1 sound level meters are often much more versatile with the possibility of measuring vibrations or inserting octave or one third octave band filters. They usually make it possible to measure a non-weighted signal (FLAT response) as well as an A-weighted and a C-weighted signal. They come with a choice from a variety of condenser microphones of different sensitivities and characteristics.

The evaluation of impulses involves the determination of the peak level and the duration of the impulse. Some precision sound level meters are equipped with a circuit that makes it possible to measure the peak level: the time constant used in this case is about 50ms and a circuit is included to hold the instantaneous level. After recording the peak value, the meter must be reset in order to read another value.

Some sound level meters offer the possibility to measure the equivalent A-weighted level $L_{Aeq,T}$ according to the equal energy principle. This can be done in two ways. In the first, the integrating period is prefixed (in some cases, 60 seconds) and the instrument computes the $L_{Aeq,T}$ level progressively: intermediary readings are then irrelevant and the user may only record the final value. In the second type, the integrating period is not fixed and the instrument actually gives the $L_{Aeq,\ T}$ level computed during the time elapsed since it was started. This type is of more use than the first one as the user does not have to define before hand the integrating time to be used.

Use of Sound Level Meters

This describes how to use physically the instrument in order to correctly measure the noise level existing at the point where the microphone is placed. The following steps must be taken successively:

1. Batteries must be checked before use and during long measuring sessions.
2. A wind shield must be used if the air velocity is noticeable. It should anyway be used all the time as a dust shield.
3. The microphone should be oriented.
4. All intruding objects such as the body of the sound level meter (SLM) or the operator itself will degrade the frequency response of the microphone at high frequencies and directivity effects will appear at much smaller frequencies. Therefore, the SLM should be, whenever possible, installed on a stable and sturdy tripod equipped with resilient blocks to isolate the sound level meter from vibration and consequent spurious readings. The operator should be at a reasonable distance (2-3 m) behind the sound level metert. Extension cables should be used if possible when measurements are to be made in a restricted area. When the instrument makes it possible, an extension rod should be used for the microphone. For walk-through surveys, the SLM should be held well away from the body.

5. The SLM must be calibrated before any measuring session using a calibrator described in Section. If the temperature of the instrument is significantly different from the ambient temperature where it will be used, it should be first warmed up before calibration and use. The calibration must be checked at the end of the session. If the instrument is not calibrated anymore, the data might have to be discarded and the reasons for this calibration change should be investigated as this might indicate an important malfunctioning of the instrument.

6. Nowadays, it is much more advantageous to use an integrating sound level meter to determine the L_{Aeq},T over a representative period of time T than to use a simple SLM on fast or slow giving an instantaneous value.

Frequency Analyzers

The objective of frequency analysis is to determine how the overall level is distributed over a range of frequencies. The most usual analysis for occupational hygiene noise studies is octave band analysis. For more detailed information, narrower bands can be used such as one-third octave analysis or constant bandwidth analysis.

A number of analyzers are available for use with the sound level meter. The simplest models are sets of passive filters (octave or one third octave) that can be inserted between the two amplifiers of the SLM. Other analyzers are specific instruments making it possible to automatically scan the whole range of frequency bands. These are sequential instruments making measurements in one band at a time. This strongly restricts their use as the noise must be constant both in amplitude and in frequency during the 5 to 10 minutes of the analysis.

More sophisticated analyzers have the possibility to make the frequency analysis in all desired bands at the same time. These are analyzers using a set of parallel filters or using the fast fourier transform of the input signal before recombining the data into the desired bands.

One important aspect to be considered about the filters is their frequency characteristics. Ideally, the filter should provide an attenuation of infinity outside the band. In practice, this is never the case. For most common filters, the attenuation at the cut off frequencies is usually around 3 dB and is some 24 dB per doubling of frequency outside that range. Figure gives the typical frequency characteristic of an octave band filter. The practical implication of this is that a signal of 100 dB at 1000 Hz for instance will give a reading of 76 dB in the octave bands centred at 500 Hz and 2000 Hz, although no energy is present at frequencies covered by these two octave bands.

The user must then be very careful when interpreting the results of the frequency analysis of a noise that includes a strong pure tone.

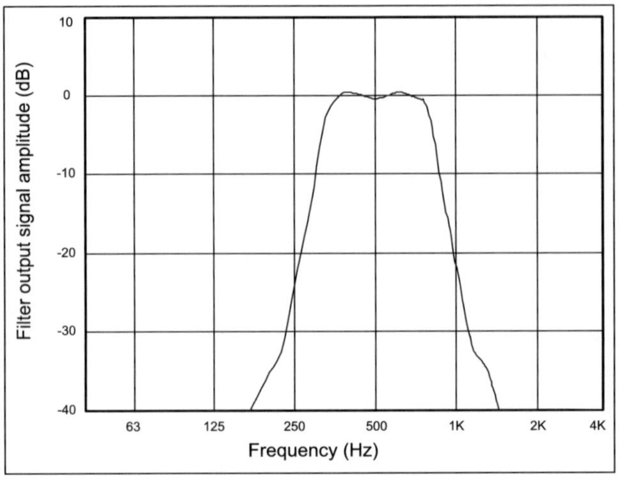

Typical 500Hz octave band filter characteristic.

As an example, consider the octave band spectrum of figure, presenting a predominant value for the 1000 Hz octave band (106 dB). A pure tone of 106 dB at 1000 Hz would give a reading of 106 - 24 = 82 dB both for the 500 Hz and the 2000 Hz octave bands. The levels of 90 and 91 dB respectively would not be very much influenced by this and therefore would reflect the total intensity at frequencies inside these bands.

However the frequency of the pure tone might be 1175 Hz: the attenuation provided by the 2000 Hz octave band filter would then be 15 dB and the level in this band 91 dB. Similarly for a 860 Hz tone, the attenuation for the 500 Hz octave band would be 16 dB and the level wrongly estimated at 90 dB.

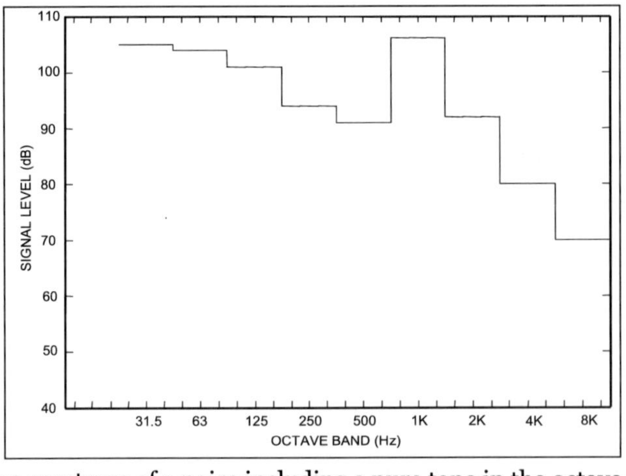

Example of the octave spectrum of a noise including a pure tone in the octave centred at 1000 Hz.

It is clear that, unless more sophisticated frequency analyses are performed, it is impossible to know precisely the frequency of the tone and therefore to determine whether the levels in the side bands are correct or not. Faced with cases like this, the user must proceed with more sophisticated analyses.

For more sophisticated instruments and especially for digital equipment, the problem is of a lesser importance as the filter attenuation with frequency is usually much higher (typically over 90 dB per octave).

Use of Frequency Analyzers

The procedure described here for carrying out a frequency analysis will only be concerned with the use of sequential octave or one-third octave filters, as with more sophisticated parallel filter equipment, the procedure might be very instrument specific. Again, only technical aspects will be presented here: the problem of deciding what type of analysis needs to be done, when and for what purpose being discussed elsewhere. Basic elements are:

1. Usually frequency analyses are performed on unweighed signals. If it is not the case, the weighting used must be clearly indicated (e.g. "A-weighted").
2. It is obvious that the noise must be fairly steady both in frequency and in amplitude during the time of the sequential analysis. If it is not steady, the noise must be recorded on a magnetic tape and the sample must be successively analyzed with each filter. Clearly this problem would not exist in equipment with parallel or FFT filters.
3. As the width of octave or one-third octave bands is smaller for low bands (31.5 and 63 Hz) than for higher ones, the fluctuations of the noise are usually much greater. Therefore the sampling time in order to get a reliable evaluation of the band level must be higher: actually, the sampling time must be inversely proportional to the width of the band (in Hz).
4. The frequency analysis should be carried on only for those bands for which the frequency characteristic of the microphone and the input amplifier is flat. Corrections may not simply be added to measured levels.
5. The filters are inserted between the input amplifier which receives the total signal and the output amplifier which must prepare the filtered signal for detection and display. For a noise such as the one represented in figure, the overall level was 113 dB, while the 8 kHz octave band level was 70 dB. The difference in this case is 43 dB which means that the output signal of the filter was 140 times (10 to the power 2.15) lower than the input signal. In such a case, care must be taken for the input signal not to saturate the input amplifier and/or the output signal not to be covered by the internal electronic noise of the output amplifier. The instruction manual of the measuring instrument should be consulted about the procedure to prevent this. Some sound level meters are provided with overload indicators for each amplifier to check that this is not the case.
6. Data for octave and one third octave analyses must be reported in a bar graph format instead of by simply joining the different points plotted at centre frequencies.

Personal Sound Level Meters

Principles

Personal sound level meters are in fact integrating sound level meters designed as dosimeters in order to be worn by the worker during his regular work. These instruments make it possible to record on almost any increment of time the equivalent level, the peak level or any statistical parameter. Typically it will record the L_{Aeq},T (in dB(A)) and L_{peak} (in dB) every second. This is extremely interesting as it makes it possible to analyse the evolution of the noise exposure during the day and to correlate it to the type of work or the location of the worker.

This type of instrument makes use of the equal energy principle and offers generally a much broader dynamic range than dosimeters. They are definitely expected to replace dosimeters in the near future and in fact are already referred to as dosimeters by some manufacturers and users. Personal sound level meters or personal sound exposure meters conform to the IEC 61252 standard.

Use

Their use is identical to the one of dosimeters with the microphone located closer to the ear of the worker.

Recorders

Graphic Level Recorder

If the sound level meter has a logarithmic DC output facility, common graphic recorders can be used to obtain a permanent record of the evolution of the sound level, providing that their writing speed is compatible with the SLOW or FAST characteristics of the SLM.

If there is no DC output or if this output is not proportional to the dB level but only to the RMS pressure, then a special recorder must be used. Many different types are available and it is not intended to review them. The essential characteristics for this type of equipment are:

- The RMS detection capabilities;
- The frequency response;
- The writing speeds, that should at least correspond to the slow and fast characteristics of the sound level meter. For reverberation time measurements, however, much faster writing speeds are needed;
- The dynamic range of the graph (often 25 or 50 dB) and of the instrument. It is usually not practical to record graphically the instantaneous noise level at a workplace for extended periods of time: the graph allows only the determination

of maximum and minimum levels and cannot be used to define any average level. The use of this technique should be restricted to special cases such as:

- The characterisation of short events of noise;
- The determination of the intermittency of a noise;
- The study of the reverberation time;
- The recording of frequency analysis.

A graphical record of the history of the L_{Aeq} noise level is usually possible with the exposure meters.

Magnetic Tape Recorders

Magnetic tape recorders are used to make a permanent recording of the noise for future analysis or reference. Some HIFI audio recorders can be used, providing their frequency response and dynamic range are suitable. For general surveys, small recorders with a frequency response of ± 3 dB in the range 30 Hz to 16 kHz and a dynamic range of 40 dB may be sufficient. For precise measurements and frequency analyses, higher quality instrumentation is needed. The real objectives of the instrument have to be assessed since the relative price of these instruments may vary in the range of 1 to 20.

As the dynamic range of an analog recorder is no more than 40 to 50 dB, usually it is difficult or impossible to record impulse noise as met in industry or as used for measuring the reverberation time. Some digital recorders (referred to as DAC recorders) are now available: they have a much broader dynamic range (around 90 dB) and a good frequency response (20 - 18000 Hz).

Besides analog and digital recorders, there are also frequency modulated (FM) recorders which are of special interest for measuring vibration as their frequency range extends down to DC.

Use of a Tape Recorder

The criteria for the selection of a tape recorder are:

- The frequency response at the different speeds. Usually the limits are directly proportional to the speed;
- The range of speeds;
- The dynamic range;
- The cross channel attenuation;

- The presence of band pass filters enabling the elimination of low frequency noise;
- The quality of the indicating device and of the input potentiometers, preferably graduated in dB;
- The possibility of controlling the output signal;
- The protection against dust;
- The protection against vibration susceptibility which increases the internal noise level.

The procedure for making a recording is as follows:

- Use preferably as the input signal, the AC output of a sound level meter, making it possible to control the level of this signal in steps of 10 dB. This will be assumed to be the case in the following steps.
- In presence of the noise, adjust the SLM so that the SLM reading is between mid and full scale. Adjust the input potentiometer of the recorder (on pause) so that the signal does not saturate the amplifier of the recorder (as indicated by the V.U. meter). The meter needle should never go into the red part of the scale. The input attenuation should not be too high, however, or the dynamic range of the recording will be reduced.
- Once this is done, do not modify any more the setting of the recorder. Record by voice the attenuation used on the SLM (for example 80 dB).
- Adjust the noise calibrator (for example, 94 dB, 1000 Hz) so that the correct level is read on the SLM. Place the SLM attenuator on the corresponding attenuation (e.g. 90 dB) and record the calibration signal for 1 or 2 minutes, mentioning clearly by voice, on the tape, what will be done and what has been done. The output signal of the SLM is in this case proportional to: attenuation setting (90) + 4 (meter reading) dB. Therefore, the recorded signal will be a reference of 84 dB, if 80 is the attenuation to be used for the real recording.
- Remove the calibrator, replace the SLM attenuator on its previous setting and after indicating orally the start of the recording and its probable duration, proceed with it. The duration of the recording is dictated by the problem to be investigated. It is well advised to make much longer recordings than at first appears needed.
- At the end of the recording, again indicate orally the stop and give full information on what was recorded, and on any particular event that might have happened during the recording.

- Repeat Step 4 of calibration.

If, for any reason, the settings of the recorder have been modified during the recording, the whole procedure must be repeated.

Calibrators

Microphones are individually calibrated at the factory, and the calibration chart must be delivered with the instrument. In the field, calibration is performed by applying a known sound pressure level at a fixed frequency to the microphone. Calibrators are small, battery driven and operate on different principles. One operates at 250 Hz and produces a sound level of 124 dB, accurate to ± 0.2 dB. To obtain the best results, the microphone should be well sealed in the coupler opening. A change in atmospheric pressure alters the calibration level slightly, but a correction can be made using the barometer which is provided as a part of the instrument set. Another example is a pocket unit, which operates at 1000 Hz. The calibration level is 94 dB with an accuracy of ± 0.5 dB.

The use of a calibrator as defined by IEC 60942 is recommended for checking the accuracy of hand-held indicating instruments, and must be used when tape recording data. Accurate calibration of equipment used in the field is essential as it provides for consistency in measurements, allows accurate comparison of measurements made over long time intervals, brings to light any slight changes in the accuracy of instrumentation, and allows a reanalysis of data, if this is required at a later date. This care in the use of calibration for field measurements should be backed up by regular laboratory calibration using more accurate techniques, in order to check the frequency response as well as the amplitude response of the equipment.

NOISE AND ITS MEASUREMENT

Sound Pressure Level

The sound pressure level is a measure of the air vibrations that make up sound. All measured sound pressures are referenced to a standard pressure that corresponds roughly to the threshold of hearing at 1,000 Hz. Thus, the sound pressure level indicates how much greater the measured sound is than this threshold of hearing. Because the human ear can detect a wide range of sound pressure levels (10–102 Pascal (Pa)), they are measured on a logarithmic scale with units of decibels (dB).

The sound pressure levels of most noises vary with time. Consequently, in calculating some measures of noise, the instantaneous pressure fluctuations must be integrated over some time interval. To approximate the integration time of our hearing system, sound pressure meters have a standard Fast response time, which corresponds to a time constant of 0.125 s. Thus, all measurements of sound pressure levels and their variation over time should be made using the Fast response time, to provide sound pressure measurements more representative of human hearing. Sound pressure meters may also include a Slow response time with a time constant of 1 s, but its sole purpose is that one can more easily estimate the average value of rapidly fluctuating levels. Many modern meters can integrate sound pressures over specified periods and provide average values. It is not recommended that the Slow response time be used when integrating sound pressure meters are available.

Because sound pressure levels are measured on a logarithmic scale they cannot be added or averaged arithmetically. For example, adding two sounds of equal pressure levels results in a total pressure level that is only 3 dB greater than each individual sound pressure level. Consequently, when two sounds are combined the resulting sound pressure level will be significantly greater than the individual sound levels only if the two sounds have similar pressure levels.

Frequency and Frequency Weighting

The unit of frequency is the Hertz (Hz), and it refers to the number of vibrations per second of the air in which the sound is propagating. For tonal sounds, frequency is associated with the perception of pitch. For example, orchestras often tune to the frequency of 440 Hz. Most environmental sounds, however, are made up of a complex mix of many different frequencies. They may or may not have discrete frequency components superimposed on noise with a broad frequency spectrum (i.e. sound with a broad range of frequencies). The audible frequency range is normally considered to range from 20–20,000 Hz. Below 20 Hz we hear individual sound pulses rather than recognizable tones. Hearing sensitivity to higher frequencies decreases with age and exposure to noise. Thus, 20,000 Hz represents an upper limit of audibility for younger listeners with unimpaired hearing.

Our hearing systems are not equally sensitive to all sound frequencies. Thus, not all frequencies are perceived as being equally loud at the same sound pressure level, and when calculating overall environmental noise ratings it is necessary to consider sounds at some frequencies as more important than those at other frequencies. Detailed frequency analyses are commonly performed with standard sets of octave or 1/3 octave bandwidth filters. Alternatively, Fast Fourier Transform techniques or other types of filters can be used to determine the relative strengths of the various frequency components making up a particular environmental noise.

Frequency weighting networks provide a simpler approach for weighting the importance of different frequency components in one single number rating. The A-weighting is most commonly used and is intended to approximate the frequency response of our hearing system. It weights lower frequencies as less important than mid- and higher-frequency sounds. C-weighting is also quite common and is a nearly flat frequency response with the extreme high and low frequencies attenuated. When no frequency analysis is possible, the difference between A-weighted and C-weighted levels gives an indication of the amount of low frequency content in the measured noise. When the sound has an obvious tonal content, a correction to account for the additional annoyance may be used.

Equivalent Continuous Sound Pressure Level

According to the equal energy principle, the effect of a combination of noise events is related to the combined sound energy of those events. Thus, measures such as the equivalent continuous sound pressure level ($L_{Aeq,T}$) sum up the total energy over some time period (T) and give a level equivalent to the average sound energy over that period. Such average levels are usually based on integration of A-weighted levels. Thus $L_{Aeq,T}$ is the average energy equivalent level of the A-weighted sound over a period T.

Individual Noise Events

It is often desired to measure the maximum level (L_{Amax}) of individual noise events. For cases such as the noise from a single passing vehicle, L_{Amax} values should be measured using the Fast response time because it will give a good correlation with the integration of loudness by our hearing system. However, for very short-duration impulsive sounds it is often desirable to measure the instantaneous peak amplitude to assess potential hearing-damage risk. If actual instantaneous pressure cannot be determined, then a time-integrated 'peak' level with a time constant of no more than 0.05 ms should be used. Such peak readings are often made using the C- (or linear) frequency weightings.

Alternatively, discrete sound events can be evaluated in terms of their A-weighted sound exposure level. The total amount of sound energy in a particular event is assessed by the SEL. One can add up the SEL values of individual events to calculate a L_{Aeq},T over some time period, T, of interest. In some cases the SEL may provide more consistent evaluations of individual noise events because they are derived from the complete history of the event and not just one maximum value. However, A-weighted SEL measurements have been shown to be inadequate for assessing the (perceived) loudness of complex impulsive sounds, such as those from large and small weapons. In contrast, C-weighted SEL values have been found useful for rating impulsive sounds such as gun shots.

Choice of Noise Measure

L_{Aeq},T should be used to measure continuing sounds such as road traffic noise, many types of industrial noises and noise from ventilation systems in buildings. When there are distinct events to the noise such as with aircraft or railway noise, measures of the individual events should be obtained (using, for example, L_{Amax} or SEL), in addition to $L_{Aeq,T}$ measurements.

In the past, time-varying environmental sound levels have also been described in terms of percentile levels. These are derived from a statistical distribution of measured sound levels over some period. For example, L10 is the A-weighted level exceeded 10% of the time. L10 values have been widely used to measure road-traffic noise, but they are usually found to be highly correlated measures of the individual events, as are L_{Amax} and SEL. L90 or L95 can be used as a measure of the general background sound pressure level that excludes the potentially confounding influence of particular local noise events.

Sound and Noise

Physically, there is no distinction between sound and noise: sound is a sensory perception evoked by physiological processes in the auditory brain. The complex pattern of sound waves is perceptually classified as "Gestalts" and are labeled as noise, music, speech, etc. Consequently, it is not possible to define noise exclusively on the basis of the physical parameters of sound. Instead, it is common practice to define noise simply as unwanted sound. However, in some situations noise may adversely affect health in the form of acoustical energy.

Measurement Issues

Measurement Objectives

The details of noise measurements must be planned to meet some relevant objective or purpose. Some typical objectives would include:

- Investigating complaints.
- Assessing the number of persons exposed.
- Compliance with regulations.
- Land use planning and environmental impact assessments.
- Evaluation of remedial measures.
- Calibration and validation of predictions.
- Research surveys.
- Trend monitoring.

The sampling procedure, measurement location, type of measurements and the choice of equipment should be in accord with the objective of the measurements.

Instrumentation

The most critical component of a sound pressure meter is the microphone, because it is difficult to produce microphones with the same precision as the other, electronic components of a pressure meter. In contrast, it is usually not difficult to produce the electronic components of a microphone with the desired sensitivity and frequency-response characteristics. Lower quality microphones will usually be less sensitive and so cannot measure very low sound pressure levels. They may also not be able to accurately measure very high sound pressure levels found closer to loud noise sources. Lower quality microphones will also have less well-defined frequency response characteristics. Such lower quality microphones may be acceptable for survey type measurements of overall A-weighted levels, but would not be preferred for more precise measurements, including detailed frequency analysis of the sounds.

Sound pressure meters will usually include both A- and C-weighting frequency-response curves. The uses of these frequency weightings were discussed above. They may also include a linear weighting. Linear weightings are not defined in standards and may in practice be limited by the response of the particular microphone being used. Instead of, or in addition to, frequencyresponse weightings, more complex sound pressure meters can also include sets of standard bandpass filters, to permit frequency analysis of sounds. For acoustical measurements, octave and one-third octave bandwidth filters are widely used with centre frequencies defined in standards.

The instantaneous sound pressures are integrated with some time constant to provide sound pressure levels. As mentioned most meters will include both Fast- and Slow-response times. Fast-response corresponds to a time constant of 0.125 s and is intended to approximate the time constant of the human hearing system. Slow-response corresponds to a time constant of 1 s and is an old concept intended to make it easier to obtain an approximate average value of fluctuating levels from simple meter readings.

Standards classify sound pressure meters as type 1 or type 2. Type 2 meters are adequate for broad band A-weighted level measurements, where extreme precision is not required and where very low sound pressure levels are not to be measured. Type 1 meters are usually much more expensive and should be used where more precise results are needed, or in cases where frequency analysis is required.

Many modern sound pressure meters can integrate sound pressure levels over some specified time period, or may include very sophisticated digital processing capabilities. Integrating meters make it possible to directly obtain accurate measures of L_{Aeq},T values over a user-specified time interval, T. By including small computers in some sound pressure meters, quite complex calculations can be performed on the measured levels and many such results can be stored for later read out. For example, some meters can determine the statistical distribution of sound pressure levels over some period, in addition to the simple L_{Aeq},T value. Recently, hand-held meters that perform loudness calculations in real time have become available. Continuing rapid developments in instrumentation capabilities are to be expected.

Measurement Locations

Where local regulations do not specify otherwise, measurements of environmental noise are usually best made close to the point of reception of the noise. For example, if there is concern about residents exposed to road traffic noise it is better to measure close to the location of the residents, rather than close to the road. If environmental noises are measured close to the source, one must then estimate the effect of sound propagation to the point of reception. Sound propagation can be quite complicated and estimates of sound pressure levels at some distance from the source will inevitably introduce further errors into the measured sound pressure levels. These errors can be avoided by measuring at locations close to the point of reception.

Measurement locations should normally be selected so that there is a clear view of the sound source and so that the propagation of the sound to the microphone is not shielded or blocked by structures that would reduce the incident sound pressure levels. For example, measurements of aircraft noise should be made on the side of the building directly exposed to the noise. The position of the measuring microphone relative to building façades or other sound-reflective surfaces is also important and will significantly influence measured sound pressure levels. If the measuring microphone is located more than several meters from reflecting surfaces, it will provide an unbiased indication of the incident sound pressure level. At the other extreme, when a measuring microphone is mounted on a sound-reflecting surface, such as a building façade, sound pressure levels will be increased by 6 dB, because the direct and reflected sound will coincide. Some standards recommend a position 2 m from the façade and an associated 3 dB correction. The effect of façade reflections must be accounted for to represent the true level of the incident sound. Thus, while locating the measuring microphone close to the point of reception is desirable, it leads to some other issues that must be considered to accurately interpret measurement results. Where exposures are measured indoors, it is necessary to measure at several positions to characterize the average sound pressure level in a room. In other situations, it may be necessary to measure at the position of the exposed person.

Sampling

Many environmental noises vary over time, such as for different times of day or from season to season. For example, road traffic noise may be considerably louder during some hours of the day but much quieter at night. Aircraft noise may vary with the season due to different numbers of aircraft operations. Although permanent noise monitoring systems are becoming common around large airports, it is usually not possible to measure sound pressure levels continuously over a long enough period of time to completely define the environmental noise exposure. In practice, measurements usually only sample some part of the total exposure. Such sampling will introduce uncertainties in the estimates of the total noise exposure.

Traffic noise studies have identified various sampling schemes that can introduce errors of 2-3 dB in estimates of daytime L_{Aeq},T values and even larger errors in nighttime sound pressure levels. These errors relate to the statistical distributions of sound pressure levels over time. Thus, the sampling errors associated with road traffic noise may be quite different from those associated with other noise, because of the quite different variations of sound pressure levels over time. It is also difficult to give general estimates of sampling errors due to seasonal variations. When making environmental noise measurements it is important that the measurement sample is representative of all of the variations in the noise in question, including variations of the source and variations in sound propagation, such as due to varying atmospheric conditions.

Calibration and Quality Assurance

Sound pressure meters can be calibrated using small calibrated sound sources. These devices are placed on the measurement microphone and produce a known sound pressure level with a specified accuracy. Such calibrations should be made at least daily, and more often if there is some possibility that handling of the sound pressure meter may have modified its sensitivity. It is also important to have a complete quality assurance plan. This should require annual calibration of all noise measuring equipment to traceable standards and should clearly specify correct measurement and operating procedures.

Source Characteristics and Sound Propagation

To make a correct assessment of noise it is important to have some appreciation of the characteristics of environmental noise sources and of how sound propagates from them. One should consider the directionality of noise sources, the variability with time and the frequency content. If these are in some way unusual, the noise may be more disturbing than expected. The most common types of environmental noise sources are directional and include: road-traffic noise, aircraft noise, train noise, industrial noise and outdoor entertainment facilities. All of these types of environmental noise are produced by multiple sources, which in many cases are moving. Thus, the characteristics of individual sources, as well as the characteristics of the combined sources, must be considered.

For example, we can consider the radiation of sound from individual vehicles, as well as from a line of vehicles on a particular road. Sound from an ideal point source (i.e. non-directional source) will spread out spherically and sound pressure levels would decrease 6 dB for each doubling of distance from the source. However, for a line of such sources, or for an integration over the complete pass-by of an individual moving source, the combined effect leads to sound that spreads cylindrically and to sound pressure levels that decrease at 3 dB per doubling of distance. Thus, there are distinct differences between the propagation of sound from an ideal point source and from moving sources. In practice one cannot adequately assess the noise from a fixed source with measurements at a single location; it is essential to measure in a number of directions from the source. If the single source is moving, it is necessary to measure over a complete pass-by, to account for sound variation with direction and time.

In most real situations this simple behaviour is considerably modified by reflections from the ground and from other nearby surfaces. One expects that when sound propagates over loose ground, such as grass, that some sound energy will be absorbed and sound pressure levels will actually decrease more rapidly with distance from the source. Although this is approximately true, the propagation of sound between sources and receivers close to the ground is much more complicated than this. The combination of direct and ground-reflected sound can combine in a complex manner which can lead to strong cancellations at some frequencies and not at others. Even at quite short source-to-receiver distances, these complex interference effects can significantly modify the propagating sound. At larger distances (approximately 100 m or more), the propagation of sound will also be significantly affected by various atmospheric conditions. Temperature and wind gradients as well as atmospheric turbulence can have large effects on more distant sound pressure levels. Temperature and wind gradients can cause propagating sound to curve either upwards or downwards, creating either areas of increased or decreased sound pressure levels at points quite distant from the source. Atmospheric turbulence can randomize sound so that the interference effects resulting from combinations of sound paths are reduced. Higher frequency sound is absorbed by air depending on the exact temperature and relative humidity of the air. Because there are many complex effects, it is not usually possible to accurately predict sound pressure levels at large distances from a source.

Using barriers or screens to block the direct path from the source to the receiver can reduce the propagation of sound. The attenuating effects of the screen are limited by sound energy that diffracts or bends around the screen. Screens are more effective at higher frequencies and when placed either close to the sound source or the receiver; they are less effective when placed far from the receiver. Although higher screens are better, in practice it is difficult to achieve more than about a 10 dB reduction. There should be no gaps in the screen and it must have an adequate mass per unit area. A long building can be an effective screen, but gaps between buildings will reduce the sound attenuation.

In some cases, it may be desirable to estimate environmental sound pressure levels using mathematical models implemented as computer programmes. Such computer programmes must first model the characteristics of the source and then estimate the propagation of the sound from the source to some receiver point. Although such prediction schemes have several advantages, there will be some uncertainty as to the accuracy of the predicted sound pressure levels. Such models are particularly useful for road traffic noise and aircraft noise, because it is possible to create data bases of information describing particular sources. For more varied types of noise, such as industrial noise, it would be necessary to first characterize the noise sources. The models then sum up the effects of multiple sources and calculate how the sound will propagate to receiver points. Techniques for estimating sound propagation are improving and the accuracy of these models is also expected to improve. These models can be particularly useful for estimating the combined effect of a large number of sources over an extended period of time. For example, aircraft noise prediction models are typically used to predict average yearly noise exposures, based on the combination of aircraft events over a complete year. Such models can be applied to predict sound pressure level contours around airports for these average yearly conditions. This is of course much less expensive than measuring at many locations over a complete one year-period. However, such models can be quite complex, and require skilled users and accurate data bases. Because environmental noise prediction models are still developing, it is advisable to confirm predictions with measurements.

Sound Transmission into and within Buildings

Sources of environmental noise are usually located outdoors; for example, road traffic, aircraft or trains. However, people exposed to these noises are often indoors, inside their home or some other building. It is, therefore, important to understand how environmental noises are transmitted into buildings. Most of the same fundamentals discussed apply to airborne sound propagation between homes in multifamily dwellings, via common walls and floors. However, within buildings we can also consider impact sound sources, such as footsteps, as well as airborne sounds.

The amount of incident sound that is transmitted through a building façade is measured in terms of the sound reduction index. The sound reduction index, or transmission loss, is defined as 10 times the logarithm of the ratio of incident-to-transmitted sound power, and it describes in decibels how much the incident sound is reduced on passing through a particular panel. This index of constructions usually increases with the frequency of the incident sound and with the mass of the construction. Thus, heavier or more massive constructions tend to have higher sound reductions. When it is not possible to achieve the desired transmission loss by increasing the mass of a panel, increased sound reduction can be achieved by a double panel construction. The two layers should be isolated with respect to vibrations and there should be sound absorbing material in the cavity. Such double panel constructions can provide much greater sound reduction than a single panel. Because sound reduction is also greater at higher frequencies most problems occur at lower frequencies, where most environmental noise sources produce relatively high sound pressure levels.

The sound reduction of buildings can be measured in standard laboratory tests, where the test panel is constructed in an opening between two reverberant test chambers. In these tests sound fields are quite diffuse in both test chambers and the sound reduction index is calculated as the difference between the average sound pressure levels in the two rooms, plus a correction involving the area of the test panel and the total sound absorption in the receiving room. The sound reduction of a complete building façade can also be measured in the field using either natural environmental noises or test signals from loudspeakers. In either case the noise, as transmitted through the façade, must be greater in level than other sounds in the receiving room. For this outdoor-to-indoor sound propagation case, the measured sound reduction index will also depend on the angle of incidence of the outdoor sound, as well as the position of the outdoor measuring microphone relative to the building façade. Corrections of up to 6 dB must be made to the sound pressure level measured outdoors, to account for the effect of reflections from the façade.

The sound reduction of most real building façades is determined by a combination of several different elements. For example, a wall might include windows, doors or some other type of element. If the sound reduction index values of each element are known, the values for the combined construction can be calculated from the area-weighted sums of the sound energy transmitted through each separate element. Although parts of the building façade, such as massive wall constructions, can be very effective barriers to sound, the sound reduction index of the complete façade is often greatly reduced by less effective elements such as windows, doors or ventilation openings. Completely open windows as such would have a sound reduction index of 0 dB. If window openings makes up 10% of the area of a wall, the sound reduction index of the combined wall and open window could not exceed 10 dB. Thus it is not enough to specify effective sound reducing façade constructions, without also solving the problem of adequate ventilation that does not compromise the sound transmission reduction by the building façade.

Sound reduction index values are measured at different frequencies and from these, single number ratings are determined. Most common are the ISO weighted sound reduction index and the equivalent ASTM sound transmission class. However, in their original form these single number ratings are only appropriate for typical indoor noises that usually do not have strong low frequency components. Thus, they are usually not appropriate single number ratings of the ability of a building façade to block typical environmental noises. More recent additions to the ISO procedure have included source spectrum corrections intended to correct approximately for other types of sources. Alternatively, the ASTMOutdoor-Indoor Transmission Class rating calculates the A-weighted level reduction to a standard environmental noise source spectrum. Within buildings the impact sound insulation index can be measured with a standard impact source and determined according to ISO and ASTM standards.

More Specialized Noise Measures

Loudness and Perceived Noise Levels

There are procedures to accurately rate the loudness of complex sounds. These usually start from a 1/3 octave spectrum of the noise. The combination of the loudness contributions of each 1/3 octave band with estimates of mutual masking effects, leads to a single overall loudness rating in sones. A similar system for rating the noisiness of sounds has also been developed. Again a 1/3 octave spectrum of the noise is required and the 1/3 octave noise levels are compared with a set of equal-noisiness contours. The individual 1/3 octave band noisiness estimates are combined to give an overall perceived noise level (PNL) that is intended to accurately estimate subjective evaluations of the same sound. The PNL metric was initially developed to rate jet aircraft noise.

PNL values will vary with time, for example when an aircraft flies by a measuring point. The effective perceived noise level measure (EPNL) is derived from PNL values and is intended to provide a complete rating of an aircraft fly-over. EPNL values add both a duration correction and a tone correction to PNL values. The duration correction ensures that longer duration events are rated as more disturbing. Similarly, noise spectra that seem to have prominent tonal components are rated as more disturbing by the tone-correction procedure. There is some evidence that these tone corrections are not always successful in improving predictions of adverse responses to noise events. EPNL values are used in the certification testing of new aircraft. These more precise measures ensure that the noise from new aircraft is rated as accurately as possible.

Aviation Noise Measures

There are many measures for evaluating the long-term average sound pressure levels from aircraft near airports. They include different frequency weightings, different summations of levels and numbers of events, as well as different time-of-day weightings. Most measures are based on either A-weighted or PNL-weighted sound pressure levels. Because of the many other large uncertainties in predicting community response to aircraft noise, there seems little justification for using the more complex PNL-weighted sound pressure levels and there is a trend to change to A-weighted measures.

Most aviation noise measures are based on an equal energy approach and hence they sum up the total energy of a number of aircraft fly-overs. However, some older measures were based on different combinations of the level of each event and the number of events. These types of measures are gradually being replaced by measures based on the equal energy hypothesis such as L_{Aeq},T values. There is also a range of time-of-day weightings incorporated into current aircraft noise measures. Night-time weightings of 6–12 dB are currently in use. Some countries also include an intermediate evening weighting.

The day-night sound pressure level L_{dn} is an L_{Aeq},T based measure with a 10 dB night-time weighting. It is based on A-weighted sound pressure levels and the equal energy principle. The noise exposure forecast (NEF) is based on the EPNL values of individual aircraft events and includes a 12 dB night-time weighting. It sums multiple events on an equal energy basis. However, the Australian variation of the NEF measure has a

6 dB evening weighting and a 6 dB night-time weighting. The German airport noise equivalent level (LEQ(FLG)) is based on A-weighted levels, but does not follow the equal energy principle.

The weighted equivalent continuous perceived noise level (WECPNL) measure proposed by ICAO is based on the equal energy principle and maximum PNL values of aircraft fly-overs. However, in Japan an approximation to this measure is used and is based on maximum A-weighted levels. The noise and number index (NNI), formerly used in the United Kingdom, was derived from maximum PNL values but was not based on the equal energy principle. An approximation to the original version of the NNI has been used in Switzerland and is based on maximum A-weighted levels of aircraft fly-overs, but its use will soon be discontinued. Changes in these measures are slow because their use is often specified in national legislation. However, several countries have changed to measures that are based on the equal energy principle and A-weighted sound pressure levels.

Impulsive Noise Measures

Impulsive sounds, such as gun shots, hammer blows, explosions of fireworks or other blasts, are sounds that significantly exceed the background sound pressure level for a very short duration. Typically each impulse lasts less than one second. Measurements with the meter set to 'Fast' response do not accurately represent impulsive sounds. Therefore the meter response time must be shorter to measure such impulse type sounds. C-weighted levels have been found useful for ratings of gun shots. Currently no mathematical description exists which unequivocally defines impulsive sounds, nor is there a universally accepted procedure for rating the additional annoyance of impulsive sounds. Future versions of ISO Standard 1996 are planned to improve this situation.

Measures of Speech Intelligibility

The intelligibility of speech depends primarily on the speech-to-noise ratio. If the level of the speech sounds are 15 dB or more above the level of the ambient noise, the speech intelligibility at 1 m distance will be close to 100%. This can be most simply rated in terms of the speech-to-noise ratio of the A-weighted speech and noise levels. Alternatively, the speech intelligibility index (formerly the articulation index) can be used if octave or 1/3 octave band spectra of the speech and noise are available.

When indoors, speech intelligibility also depends on the acoustical properties of the space. The acoustical properties of spaces have for many years been rated in terms of reverberation times. The reverberation time is approximately the time it takes for a sound in a room to decrease to inaudibility after the source has been stopped. Optimum reverberation times for speech have been specified as a function of the size of the room. In large rooms, such as lecture halls and theaters, a reverberation time for speech of about 1 s is recommended. In smaller rooms such as classrooms, the recommended value for speech is about 0.6 s. More modern measures of room acoustics have been found to be better correlates of speech intelligibility, and some combine an assessment of both the speech/noise ratio and room acoustics. The most widely known is the speech transmission index (STI), or the abbreviated version of this measure referred to as RASTI. In smaller rooms, such as school classrooms, the conventional approach of requiring adequately low ambient noise levels, as well as some optimum reverberation time, is probably adequate to ensure good speech intelligibility. In larger rooms and other more specialized situations, use of the more modern measures may be helpful.

Indoor Noise Ratings

The simplest procedure for rating levels of indoor noise is to measure them in terms of integrated A-weighted sound pressure levels, as measured by L_{Aeq},T. This approach has been criticized as not being the most accurate rating of the negative effects of various types of noises, and is thought to be particularly inadequate when there are strong low-frequency components. Several more complex rating schemes are available based on octave band measurements of indoor noises. In Europe the noise rating system, and in North America the noise criterion, both include sets of equal-disturbance type contours. Measured octave band sound pressure levels are compared with these contours and an overall noise rating is determined. More recently, two new schemes have been proposed: the balanced noise criterion procedure and the room criterion system. These schemes are based on a wider range of octave bands extending from 16–8,000 Hz. They provide both a numerical and a letter rating of the noise. The numerical part indicates the level of the central frequencies important for speech communication and the letter indicates whether the quality of the sound is predominantly low-, medium- or high-frequency in nature. Extensive comparisons of these room noise rating procedures have yet to be performed. Because the newer measures include a wider range of frequencies, they can better assess a wider range of noise problems.

SOUND INTENSITY

Sound intensity is the amount of energy Sound intensity, amount of energy flowing per unit time through a unit area that is perpendicular to the direction in which the sound waves are travelling. Sound intensity may be measured in units of energy or work—e.g., microjoules (10^{-6} joule) per second per square centimetre—or in units of power, as microwatts (10^{-6} watt) per square centimetre. Unlike loudness, sound intensity is objective and can be measured by auditory equipment independent of an observer's hearing.

The intensity of one sound can be compared to that of another of the same frequency by taking the ratio of their powers. When this ratio is 10, the difference in intensity of the sounds is said to be one bel, a unit named in honour of the United States inventor Alexander Graham Bell. Accordingly, the relative intensities of two sounds in bels is equal to the logarithm of the intensity ratio—i.e., if I is the intensity of one sound and I_0 is that of another, then the intensity ratio B in bels is $B = \log_{10} (I/I_0)$. The unit in general use is the decibel (abbreviated db), equal to 0.1 bel. Thus the equation for relative intensities b in decibels may be written $b = 10 \log_{10} (I/I_0)$. It may be calculated from this equation that one decibel corresponds to a 26 percent change in intensity. If I_0 is taken as a reference standard intensity equal to 10^{-16} watt per square centimetre, the intensity of the faintest sound that can be heard, then the intensity, or level, of any sound can be measured in decibels. Thus speech of intensity 10^{-12} watt per square centimetre has a sound level of $b = 10 \log_{10} (10^{-12}/10^{-16}) = 40$ decibels.

NOISE CALCULATION

Noise calculation is the process of calculating the level of noise immission using the metric dB(A). Noise immission is created by noise sources (noise emission) of various types which are propagating noise into the environment. A single source will create a certain level of immission primarily driven by originating sound power level and distance influenced by e.g. absorption and reflection. Several noise sources result in typically higher levels of immission. The method or process of determining the resulting immission level is called noise calculation, its graphical representation is called noise map.

Noise calculation is done for the following reasons:

- To calculate a current situation based on existing physics, mostly when a physical measurement at a location is impractical;
- To predict the resulting immission levels based on a planned change, e.g. set up of a new machine.

Mostly, noise calculation is part of any such planning process and may become part of the decision process for physical change.

Environment engineering differentiates between:

- Indoor sound propagation, e.g. occupational health considerations;
- Outdoor sound propagation, e.g. city planning considerations as well as type of noise source, e.g. Industrial noise, rail or air traffic noise, road noise.

Each of before mentioned require different algorithms for noise calculation which have been defined over time in a number of different national or international standards. As an example, outdoor sound propagation is defined in ISO standard 9613 which can be received from the ISO organisation or scientific literature.

Calculation Standards

This list of widely accepted noise calculation standards is incomplete, and new standards may be under development:

- ISO 9613 parts 1 and 2' defines the calculation methods for outdoor sound propagation (International).
- VDI2714 and VDI2720' Schallausbreitung und Schallschutz im Freien (outdoor sound propagation, Germany).
- DIN18005' Schallschutz im Städtebau (noise protection in cities, Germany).
- TA Laerm' Technische Anleitung zum Schutz gegen Lärm (Germany).
- VDI-Richtlinie 3745 Blatt 1' Beurteilung von Schießgeräuschimmissionen (Germany).
- BS 5228' Code of practice for noise and vibration control on construction and open sites (UK).
- RLS-90' Richtlinien für den Lärmschutz an Straßen (noise protection at roads/ traffic, Germany).
- CRTN' calculation of road traffic noise (UK).
- Schall03' Schienenlärm (calculation of rail noise, Germany).

Calculation Tools

Noise experts and some small specialized companies have slowly developed a limited number of calculation tools, which have increased in number and become more user-friendly, covering more application cases, and adding service elements to the noise calculation tools. The noise calculation process is complex in input (gathering data, correctly modeling acoustic elements in the field) and from a pure compute power perspective, recently advances involved:

- Moving calculation routines into self-contained calculation engines;
- Adapting CAD systems for input of acoustic models;
- Moving into 3D for displaying models or even data capture;
- Moving into cloud services and web enabled calculation tools.

The quality of results from the noise calculation depends on the quality of the acoustic data provided and the capability of the noise calculation engine.

Some more widely used noise calculation software packages are:

- Noise3d online;
- Code_Tympan;
- IMMI;
- Olive Tree Lab Terrain.

AMBIENT NOISE LEVEL

In atmospheric sounding and noise pollution, ambient noise level (sometimes called background noise level, reference sound level, or room noise level) is the background sound pressure level at a given location, normally specified as a reference level to study a new intrusive sound source.

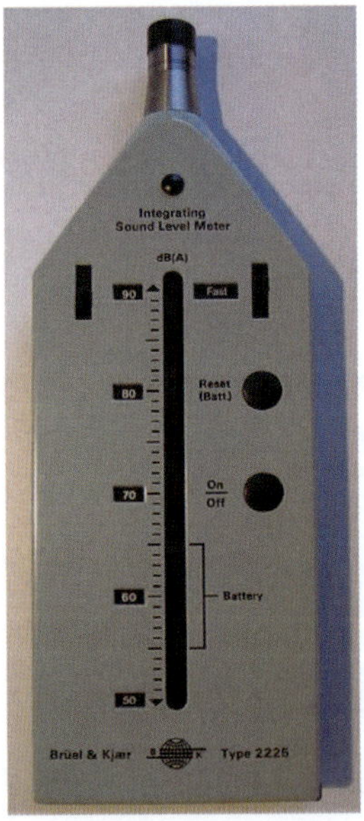

Integrating Sound Level Meter.

Ambient sound levels are often measured in order to map sound conditions over a spatial regime to understand their variation with locale. In this case the product of the investigation is a sound level contour map. Alternatively ambient noise levels may be measured to provide a reference point for analyzing an intrusive sound to a given environment. For example, sometimes aircraft noise is studied by measuring ambient sound without presence of any overflights, and then studying the noise addition by measurement or computer simulation of overflight events. Or roadway noise is measured as ambient sound, prior to introducing a hypothetical noise barrier intended to reduce that ambient noise level.

Ambient noise level is measured with a sound level meter. It is usually measured in dB relative to a reference pressure of 0.00002 Pa, *i.e.,* 20 μPa (micropascals) in SI units. A pascal is a newton per square meter. The centimeter-gram-second system of units, the reference sound pressure for measuring ambient noise level is 0.0002 dyn/cm^2. Most frequently ambient noise levels are measured using a frequency weighting filter, the most common being the A-weighting scale, such that resulting measurements are denoted dB(A), or decibels on the A-weighting scale.

NOISE DOSIMETER

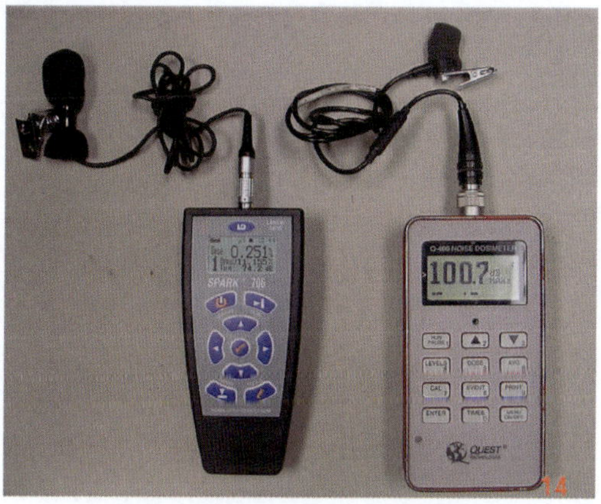

Personal noise dosimeters.

A noise dosimeter (American English) or noise dosemeter (British English) is a specialized sound level meter intended specifically to measure the noise exposure of a person integrated over a period of time; usually to comply with Health and Safety regulations such as the Occupational Safety and Health (OSHA) 29 CFR 1910.95 Occupational Noise Exposure Standard or EU Directive 2003/10/EC.

Noise Measurement

Noise dosimeters measure and store sound pressure levels (SPL) and, by integrating these measurements over time, provide a cumulative noise-exposure reading for a given period of time, such as an 8-hour workday. Dosimeters can function as personal or area noise monitors. In occupational settings, personal noise dosimeters are often worn on the body of a worker with the microphone mounted on the middle-top of the person's most exposed shoulder.

Area monitoring can be used to estimate noise exposure when the noise levels are relatively constant and employees are not mobile. In workplaces where employees move about in different areas or where the noise intensity tends to fluctuate over time, noise exposure is generally more accurately estimated by the personal monitoring approach.

Worker wearing a personal noise dosimeter.

Dosimeters are also used to collect data for use in legal proceedings, development of engineering noise controls, and other industrial hygiene purposes. When planning to conduct noise exposure measurements, steps must be taken to ensure that the dosimeters are calibrated and operated according to manufacturers' specifications. It is also necessary to understand the properties of the acoustic environment, the main measurement objectives as they relate to determining the risk to hearing damage, and the limitations associated with the use of dosimeters. Dosimeter manufacturers recommend that the instrument be calibrated with an acoustical calibrator such as a pistonphone before and after each measurement to verify reliable operation. In addition to field calibration routines, the manufacturers recommend periodic comprehensive calibration and certification of the instrument by an accredited laboratory using traceable reference sources. Field calibration of contemporary dosimeters has been mostly automated through PC-based programs that run the calibration routine, document the time and date, and adjust for any offset in levels.

Occupational Settings

Current dosimeters are designed to provide the user with parameters such as noise dose, time-weighted average, sound exposure level, as well as peak, maximum, and minimum sound pressure levels. Most dosimeters also generate statistical and graphical representations of the collected data. ANSI S1.25 specifies that dosimeters should at least provide the following parameters:

- Frequency weighting: A-weighting or C-weighting;
- Exponential averaging: F (fast); S (slow);

- Criterion level: 90, 85, 84, 80, or V (variable);
- Criterion duration: Hours;
- Threshold level: 90, 80, or V (variable);
- Exchange rate: 5, 4, or 3.

A noise or sound dose is the amount of sound a person is exposed to in a day. The dose is represented by a percentage. A noise dose of 100% means that a person has exceeded the permissible amount of noise. Any noise exposure after the 100% noise dose may damage hearing. The exchange rate is the rate at which exposure accumulates. An addition of the exchange rate results in a halving of exposure time. The following table represents the exposure levels for occupational organizations as of November 2018:

OSHA	NIOSH	FRA	MSHA
Permissible Exposure Level: 90 dBA 8-hour TWA Action Level: 85 dBA 8-hour TWA	Recommended Exposure Limit: 85 dBA 8-hour TWA	Permissible Exposure Level: 90 dBA 8-hour TWA Action Level: 85 dBA 8-hour TWA	Permissible Exposure Level: 90 dBA 8-hour TWA Action Level: 85 dBA 8-hour TWA
5 dB exchange rate	3 dB exchange rate	5 dB exchange	5 dB exchange rate

Standardization

The international body that specifies the technical requirements of such instruments as sound level meters and dosimeters is the International Electrotechnical Commission (IEC) based in Geneva; whereas the method of their use is normally given in an International Organization for Standardization (ISO) publication. In the U.S., the American National Standards Institute (ANSI) ANSI S1.25-1991 (R2007) specifies the performance characteristics of personal noise dosimeters.

Use of Dosimeters

The original dosimeters were designed to be belt worn with a microphone connected to the body of the dosimeter and mounted on the shoulder as near to the ear as practicable. These devices were worn for the full work shift and at the end would give a readout initially in percentage dose, or in some other exposure metric. These were the most common way of making measurements to meet legislation in the USA, but in Europe, the conventional sound level meter was favoured. There were many reasons for this, but in general in Europe the dosimeter was distrusted for several reasons, some being:

- The cable was considered dangerous as it could catch on rotating machinery;

- The dosimeter could tell you the level had been exceeded, but it did not say when this happened;
- Workers could falsify the data very easily;
- The device was big enough to affect the work pattern;
- In the USA - where most of the early devices were manufactured, these reasons did not seem to matter so much.

To remove these European objections, dosimeters became smaller and started to include a data store where the Time History of the noise, usually in the form of Short Equivalent Sound Level (Leq) was stored. This data could be transferred to a personal computer and the exact pattern of the noise exposure minute by minute plotted. The usual method used was to store data in the form of Short Leq, a French concept that helped to bring computers into acoustics. As well, dosimeters started to incorporate besides the A-weighting a second C-frequency-weighted channel that allowed the true peak to be indicated. By the time the PSEM standard was published, many major sound level meter companies - in both Europe and the USA had a dosimeter in their range.

Noise dosimeters are worn by workers in order to track their sound exposure over a period of time. With the accuracy of a type 2 sound level meter, a majority of noise dosimeters measure within ±2 dB A. One must make sure to the noise dosimeter is properly calibrated and kept out of extreme temperature and humidity. The noise dosimeter is typically programmed by a hearing conservationist, sound engineer or audiologist. When the professional is setting up the noise dosimeter, settings like frequency of sound sampling and log information should be considered. When placing any dosimeter, the microphones should be clipped to the shoulder with the microphone facing upwards. The microphone should be placed in the open and clear from any surrounding fabric. It should also be protected from any wind source when outdoors and should have a wind screen over it for protection if needed. Over the course of the day, the dosimeter will measure the time-weighted average of the sound level the user experienced. It is important to make sure the batteries are fully charged since the dosimeter often has to run for 8 to 10 hours over the course of the work shift. Noise dosimeters do not record the user's voice, so it is important to teach the user how to operate and successfully use the device so the end result is not influenced by the user's tampering.

SOUND EXPOSURE

Sound Exposure (E) is the (sound-pressure)^2 measured over a stated period of time or noise event. Measurements are normally A-weighted to relate to the human response.

The SI unit of sound exposure is the $Pa^2 \cdot s$ (pascal-squared second), however Occupational Noise Exposure Meters, also known as noise dose meters, record the $Pa^2 \cdot h$ (pascal-squared hour) levels throughout over the working day.

For a given period of time, an increase of 10 dB(A) in sound pressure level corresponds to a tenfold increase in the sound exposure.

- 1 $Pa^2 \cdot h$ = 100% Dose = 85 L_{Aeq}(8 h) = 85 dBA for 8 hours.
- 0.1 Pa^2 = 74.9 L_{Aeq},8 h.
- 0.2 Pa^2 = 78.0 L_{Aeq},8 h.
- 0.5 Pa^2 = 81.9 L_{Aeq},8 h.
- 1 Pa^2 = 84.9 L_{Aeq},8 h = 100% Dose.
- 2 Pa^2 = 88.0 L_{Aeq},8 h.
- 5 Pa^2 = 91.9 L_{Aeq},8 h.
- 10 Pa^2 = 94.8 L_{Aeq},8 h.

Sound Exposure Action Values (EAV) is the 8-hour daily exposure to noise above which employers are required to take action to control exposure. For noise there are two action levels:

- The Lower EAV is 80 dBA and a peak sound pressure of 135 dBC ≈ LCpeak.
- The Upper EAV is 85 dBA and a peak sound pressure of 137 dBC ≈ LCpeak.

Sound Exposure Limit Value (ELV) is the maximum noise an employee may be exposed to on any single 8-hour day is 87 dBA and a peak sound pressure of 140 dBC.

Sound exposure is the time integral of the squared A-weighted, instantaneous sound pressure, over a stated period of time or event. The frequency weighting may be other than A, if so specified.

Most modern sound level meters measure sound exposure and then calculate the required parameters.

Sound Exposure Level (LE) is the constant sound level that has the same amount of energy in one second as the original noise event. A-weighted sound exposure levels are denoted by the symbol LAE.

Sound Exposure Level is similar to the Leq - equivalent continuous sound level as the total sound energy is integrated over the measurement period. However instead of averaging over the measurement period, a reference duration of 1 second is used.

It follows that the sound exposure level = Leq + 10·Log10T where T is in seconds for the whole measurement period.

Sound Exposure Level (SEL) is numerically equivalent to the total sound energy. For example a noise level of 90 dBA lasting 1 second would have a SEL of 90 dBA but if the event lasted 2 seconds the SEL would be 93 dBA. Put another way if a second event of 80 dBA occurred it would have to last 10 seconds to register a 90 dBA SEL.

Sound Exposure Levels normalized to 1 second are a very useful way of comparing different sound events and sources.

Logarithm of the ratio of a given time integral of squared A-frequency weighted sound pressure, over a stated time interval or event such as an aircraft flyover, to the product of the squared reference sound pressure of 20 μPa and the reference duration of one second. Sound exposure level in decibels is ten times the logarithm to the base ten of that ratio. Sound Exposure Meter a small instrument designed to be worn by an individual to provide a measure of the accumulated sound exposure received by the wearer while moving about during the workday.

The instrument is calibrated in $Pa^2 \cdot h$. If the meter is worn for only a representative part of the working day, the reading must be corrected appropriately.

METRICS FOR ASSESSING ENVIRONMENTAL NOISE

Selecting a metric for assessment of environmental noise is no simple task, because it must reflect the impact on people. No single metric can describe all responses in all situations. Context, expectations, and people's experiences and circumstances all affect their responses. Hence, levels of community response (e.g., annoyance) may vary from community to community, just as individual responses vary from person to person, even if noise levels do not change. However, one consistent finding has been that changes in noise exposure do affect individual and community responses and that increases in man-made noise usually have a negative impact.

Thus, it is important to understand which characteristics of noise elicit a negative response and how exposure to noise with those characteristics affects people's lives. The metric chosen or developed for measuring community noise must reflect this human response and must be taken into account in making policy decisions.

Fifty years ago, when noise metrics were developed, the choices were based on simpler calculations and technologies and the acoustical quantities that could be predicted by sound propagation models used at the time. Although much more sophisticated measurements can be made today, many still consider these "older" metrics valid and continue to use them. However, with modern instruments, much more accurate measurements and predictions can now be made of people's reactions to noise.

A meaningful metric, or set of metrics, translates sound pressure-time history measurements into a prediction of the effects of noise, such as annoyance, sleep disturbance, changes in health, interference with understanding speech, and ability to learn. Ideally, this translation should be based on context, expectations, and personal situations and preferences, in addition to noise information, and should account for a distribution of responses, including responses of vulnerable populations, such as children. Unfortunately, a holistic model of community response is still beyond present capabilities.

One fundamental issue that must be considered in the choice of an environmental noise metrics is the purpose for which the metric will be used:

- To implement public policy on noise immission from one or more sources;
- To provide information about noise exposures in a form understandable to the public;
- To assess a noise situation in terms of noise control engineering.

The metrics to accomplish these purposes may differ, but all three relate directly to the impact of noise on the community. For example, a metric to inform decisions about noise control engineering strategies should result in reducing the noise impact, which would then be reflected in the policy metrics and the public information metrics.

As new research results become available and accessible, they should influence the choice of metrics for the three purposes listed. The results of such research may result in complex calculations that include many variables and may better quantify individual reactions to sound. Some modern procedures, such as calculation of loudness, are more complex than earlier methods, but available computational procedures make the results widely available.

In addition, alternative metrics are described that may be easier for the public to understand than the day-night average sound level (DNL).

Loudness and A-Weighting

Arguably the modern history of noise metrics began in the 1930s with the search for a way to describe the loudness of sound. This led to the definition of weighting networks for sound-level meters and, because of limitations on the capabilities of calculating sound pressure levels at that time, a single frequency-weighted value—either A-weighted or C-weighted—came into common usage.

Loudness

In an early attempt to determine the loudness of sound (using discrete-frequency tones), Fletcher and Munson found that the loudness of a tone depends on both its amplitude and its frequency. Knowing this dependence, they were able to develop a set of equal-loudness curves. In modern terms the unit of loudness is the phon. For example, a 1,000-Hz tone with a sound pressure level of 40 dB has a loudness of 40 phon. At this loudness level the sound pressure level of tones between 1,000 and about 5,000 Hz is generally lower than 40 dB, and the sound pressure level of tones below 1,000 Hz and above about 5,000 Hz is higher than 40 dB.

The sound-level meter was standardized in the early 1930s when microphones and electronic circuits were being developed. Ideally, the standard sound-level meter would have a single-number description of the sound at a given point in space. The best description at the time came from the studies by Fletcher and Munson, who clearly showed that the shape of the equal-loudness curve was dependent on both the amplitude and the frequency of sound. Thus, using the linear electronic circuits of the time, a few curves had to be selected based on the amplitude of the sound. One of the curves selected, which is very close to the 40-phon curve, was designated as "A-weighting." Another, which was nearly independent of frequency, was designated as "C-weighting." A third curve, the "B-weighting" curve, which fell between the A and C curves, has long since fallen out of favor. A-weighting and C-weighting are still used today, although the shape of the curves has changed somewhat to provide a standardized mathematical description in terms of poles and zeros of a transmission network.

Work on improving the calculation of loudness based on measurement of the spectrum of sound continued. The best-known early work in the United States was by S. S. Stevens and in Germany by Eberhard Zwicker. Stevens's Mark VI and Zwicker's work on loudness were standardized by the International Organization for Standardization. Later work by Brian Glasberg and Brian Moore in the United Kingdom was the basis for the American National Standard on computation of loudness.

Over the years, A-weighted levels were found to correspond reasonably well to human response, especially for noise spectra in typical offices. Single-number methods of rating noise in offices and other building spaces were also developed, including so-called noise rating curves (NR curves—a curve tangent method of obtaining a single number from an octave band spectrum) and ratings based on loudness and A-weighting.

Metrics for Measuring Community Reaction to Noise

One early attempt to develop a metric for forecasting community response to noise was made by Stevens et al. Unlike the DNL, this metric included nonacoustical factors as well as noise levels and yielded a "composite noise rating." This rating was then plotted against a scale of community responses—vigorous community action, threats of community action, widespread complaints, sporadic complaints, and no observed reaction. A few case studies showed a reasonable correlation between the measurement and response but with considerable scatter. Community noise levels were determined by measuring the average octave band levels in the community averaged in space and time. A curve tangent method was used to reduce the octave band data to a single-number rating.

Day-night Average Sound Level

After EPA established the Office of Noise Abatement and Control and after passage of the Noise Control Act of 1972, EPA was faced with the task of developing a metric for community noise with the following characteristics:

- The measure should be applicable to the evaluation of pervasive long-term noise in various defined areas and under various conditions over long periods of time.
- The measure should correlate well with known effects of the noise environment on the individual and the public.
- The measure should be simple, practical, and accurate. In principle, it should be useful for planning as well as for enforcement or monitoring purposes.
- The required measurement equipment, with standardized characteristics, should be commercially available.

  Weighting is less useful for measuring human response to sound when the spectrum has a large low-frequency component, when high-amplitude peaks in the spectrum are in the 2- to 4-kHz range, and when the sound is tonal or impulsive.

- The measure should be closely related to existing methods currently in use.
- The single measure of noise at a given location should be predictable, within an acceptable tolerance, from knowledge of the physical events producing the noise.
- The measure should lend itself to small, simple monitors that can be left unattended in public areas for long periods of time.

With respect to both simplicity and adequacy for characterizing human response, a frequency-weighted sound level should be used for the evaluation of environmental noise. Several frequency weightings have been proposed for generalusein the assessment of response to noise, differing primarily in the way sounds at frequencies between 1000 and 4000 Hz are evaluated. The A-weighting, standardized in current sound level meter specifications, has been widely used for transportation and community noise description. For many noises, the A-weighted sound level has been found to correlate as well with human response as more complex measures, such as the calculated perceived noise level or the loudness level derived from spectral analysis. However, psycho-acoustic research indicates that, at least for some noise signals, a different frequency weighting which increases the sensitivity to the 1000–4000 Hz region is more reliable. Various forms of this alternative weighting function have been proposed; they will be referred to here as the type "D-weightings." None of these alternative weightings [have] progressed in acceptance to the point where a standard has been approved for commercially available instrumentation.

It is concluded that a frequency-weighted sound pressure level is the most reasonable choice for describing the magnitude of environmental noise. In order to use available standardized instrumentation for direct measurement, the A-frequency weighting is the only suitable choice at this time. The indication that a type D-weighting might ultimately be more suitable than the A-weighting for evaluating the integrated effects of noise on people suggests that at such time as a type D-weighting becomes standardized and available in commercial instrumentation, its value as the weighting for environmental noise should be considered to determine if a change from the A-weighting is warranted.

The decision to add 10 dB in measuring nighttime levels and the selection of a two-period (day-night) metric rather than a three-period metric (day-evening-night) was based on community reaction studies at the time and tests that showed little difference between a two-period and a three-period metric. Thus, the DNL (A-frequency weighting for both day-time and nighttime levels and a 10-dB increase in measuring system gain at night) came into being for the evaluation of community noise.

In the United States, DNL and the percentage of persons highly annoyed are widely used, especially by the Federal Aviation Administration (FAA). The Federal Highway Administration uses A-weighting and the average sound pressure level during the busiest traffic hour as a measure of community impact. The difference between C-weighted and A-weighted levels is used as an indication of the low-frequency content of the sound, and the sound exposure level is used to evaluate sounds of finite duration—for example, an aircraft flyover.

Day-evening-night sound level is widely used in Europe. In some countries, L_{day} and L_{night}, (average A-weighted sound pressure levels) are used in addition to or instead of a DNL-type metric. None of these metrics takes into account the time of night when the noise occurs, even though noise appears to cause greater sleep disturbance at the beginning and end of the night.

Several issues have arisen from the use of DNL and the percentage of persons highly annoyed: no one actually "hears" a DNL; there is a high variability from study to study around a nominal Schultz curve; and in many situations "highly annoyed" is not an appropriate measure of human response. Although the percent highly annoyed and DNL approach has been widely endorsed, variability around a nominal Schultz curve is troubling, and there are reports that this approach is not sufficient to predict community response. Attitudinal and personal variables impact people's responses and are, to some extent, the reason for scatter.

Some researchers have found in their analyses of survey results that the nominal Schultz curve appears to depend on the noise source (e.g., aircraft, road traffic, rail traffic). In addition, DNL is a relatively insensitive measure of sleep disturbance and thus is not an appropriate metric for predicting awakenings in sleep disturbance studies. Finally, A-weighting is not the best weighting for measuring noises with unusual spectra (e.g., excessive high- or low-frequency noise or noise that has unusual peaks in its spectrum). For sounds with levels that evolve over time, the most appropriate.

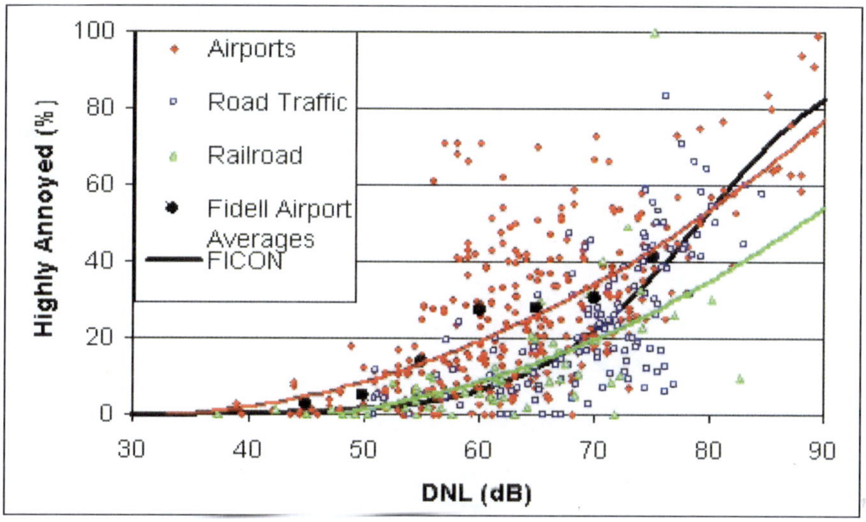

Curves are the results of fits to data associated with different modes of transportation.

A number of metrics have been developed to take into account day-time versus night-time operations around airports. These include Noise Exposure Forecast, Community Noise Equivalent Level, and Noise and Number Index. The EPA rationale for selecting a 10-dB nighttime penalty is as follows: "Methods for accounting for the differences in interference or annoyance between daytime/nighttime exposures have been employed in a number of different noise assessment methods around the world. The weightings applied to the nondaytime periods differ slightly among the different countries but most of them weight night activities on the order of 10 dB; the evening weighting if used is 5 dB. The choice of 10 dB for the nighttime weighting was predicated on its extensive prior usage, together with an examination of the diurnal variation in environmental noise."

Weighting should change with the level; typically, however, only one weighting is used.

Percentage of Persons Highly Annoyed

The next major event in the selection of a noise metric was a study by Schultz of surveys of community reaction to noise. Schultz went back to original data to estimate the percentage of the population "highly annoyed" as a function of DNL. Even at that time, it was recognized that, for a variety of reasons, there was considerable scatter in the data. Nevertheless, Schultz proposed that a single curve (the Schultz curve) drawn through the data should be used as a measure of community response. Later studies led to modifications of the Schultz curve. In the latter study, three curves were compared, and a U.S. Air Force logistic curve was defined,

$$\%\mathrm{HA} = 100/[1+\exp(11.13-0.14L_{dn})]$$

The scatter in the highly annoyed response, compared to scatter in the average curve, was presented by Miedema and Vos and has been commented on by several subsequent researchers. The first problem with scatter is that it causes great uncertainty in the prediction of community reaction. A second problem is that community reaction (percent highly annoyed) appears to depend on the source of the noise; for example, responses to aircraft noise, road traffic noise, and rail noise vary, even if the noises have the same DNL. The question that must be answered is whether the variability in response is due to the nature of the noise source or reflects how the metric is calculated.

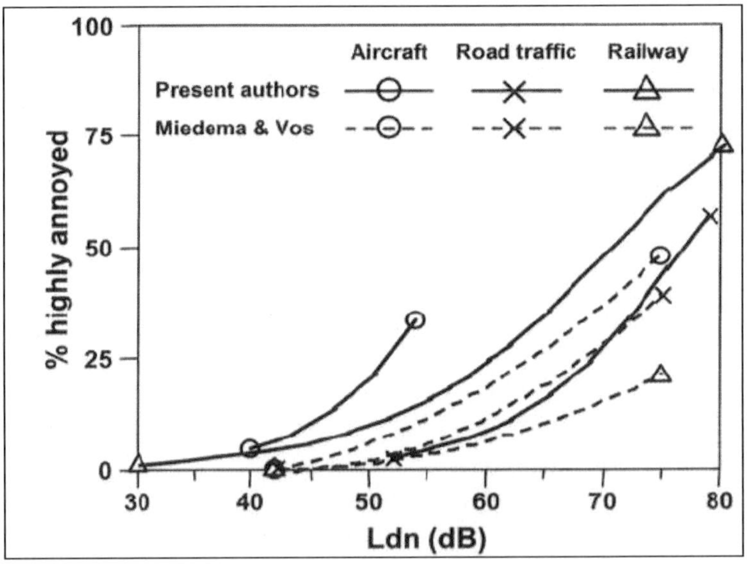

Consultants and other professionals are often asked to study community noise issues and recommend remedial action. Predictions of community response should not be based only on variations of the Schultz curve. It has been known for many years that nonacoustical factors influence community reaction to noise. Thus, at a minimum, temporal and spectral variations must also be taken into account.

Based on work by EPA, Schomer proposed modifications to DNL to account for tonality, impulsiveness, background noise, type of community, and other factors. Schomer also showed how this modified approach could be used to reduce variances in the survey data on which the Schultz curve is based.

The Federal Interagency Committee on Noise endorsed the use of percent highly annoyed and DNL as metrics for assessing community noise around airports and recommended that the equation be accepted as showing the definitive relationship between percent highly annoyed and DNL. Response curves for community annoyance have now been standardized nationally and internationally.

Alternative Metrics

The science of measuring environmental noise has progressed rapidly in the past decade as computer technology has come on line to provide rapid data acquisition and analysis in small portable packages. The end result has been a revolution in the type and complexity of measurements and calculations that can be made in analyzing environmental noise.

Different Frequency Weighting

An alternative to A-weighting (i.e., D-weighting) could be considered. As noted earlier, this weighting was considered by the EPA in 1974 but rejected because there was no standard shape for the curve.

Perceived Noise Level

Community reaction to noise from jet planes led to important events in the development of noise metrics. The problem, which arose in 1956, is described in an autobiography by Beranek. According to measurements made with a standard sound-level meter, the noise produced by a Boeing 707 jet airplane and that by a propeller airplane (Super Constellation) were equal. However, subjective testing showed that the 707 was considered much noisier; by subjective measures, the A-weighted sound pressure levels of the 707 would have to be significantly reduced to be considered as noisy as the Super Constellation. This early test of the usefulness of A-weighted levels in judging noisiness led to further evaluations of the relative noisiness of propeller-driven and jet airplanes and the development of the concept of "perceived noisiness".

Perceived noise level (PNL) was used in the development of specifications of noise emissions from airplanes for regulatory purposes in 1969 and is still used to certify airplanes today. When the perceived noise-level metric was adopted, it was possible to compute it only with a large amount of equipment. Today, it can be done with a handheld sound-level meter. D-weighting simplifies the PNL calculation, but neither PNL nor D-weighting solves the decibel issue, which relates to explaining noise to the public.

Loudness

Historically, the method of calculating PNL was similar to the method of calculating loudness. Today, several methods can be used to calculate loudness, all of them with a handheld sound-level meter. Loudness that exceeds some agreed-on value a given percentage of the time also can be calculated. On a linear scale (as opposed to a logarithmic scale), a doubling of the value of the calculated value corresponds to a doubling of the loudness. This may be easier to explain to the public than a metric that uses the phon (which uses a logarithmic scale) as a unit of loudness. For sounds in the midfrequency range, an increase in A-weighted level of 10 dB corresponds to a doubling of loudness.

Speech Interference

Standard methods of calculating speech interference are available, and the values may be translated into effects that are easier for the public to understand than DNL. For example, the difficulty of communicating over a given distance between speaker and listener may be quantified in terms of percentage of speech likely to be understood. Speech interference can be affected by the fact that hearing loss increases with age and usually starts at high frequencies. Thus, the ability to distinguish consonants that have high-frequency content such as "s" and "th" is diminished.

Noise Metrics for Rural/Naturally Quiet Areas

Neither day-night average sound level nor percent highly annoyed is an appropriate metric for measuring noise in naturally quiet areas. Because of the logarithmic nature of the decibel, short-duration sounds of high amplitude compared with background noise can significantly increase the day-night level, even though the sound remains at the background level most of the time. As for percent highly annoyed, this is hardly the best measure of satisfaction for areas where quiet and solitude are valued. In addition, it can be difficult to measure very low sound pressure levels. A-weighted levels of 40 dB are at the upper end of the range, and lower levels can be at or even below the levels measurable with conventional sound-level meters.

Nevertheless, some quantification of noise impact is clearly needed in these areas as a basis for establishing public policy, which usually means regulatory action. The classic definition of noise is "unwanted sound," so the source of sound must be identified, either as part of the natural soundscape or not. Thus, simple metrics like sound pressure level are clearly not appropriate. For example, an airplane over flight may have a much lower sound pressure level and shorter duration than sound from a rushing stream, but the former is considered noise and the latter is considered sound. The method of assessment of the noise environment should also take into account the likely long-term impact on animals that use, for example, very low level sounds (perhaps inaudible or unnoticed by people) to locate prey or predators.

References

- Kardous, chucri a. (8 april 2008). Noise dosimeters. Handbook of noise and vibration control. Pp. 465–469. Doi:10.1002/9780470209707.ch39. Isbn 9780470209707

- Sound-intensity, science: britannica.com, Retrieved 25 February, 2019
- Www. Noiseandthecity.org (2017-10-02). "noisescore—a mobile app for community noise issues". Noiseandthecity.org. Retrieved 2019-01-12
- Sound-exposure: acoustic-glossary.co.uk, Retrieved 16 January, 2019
- "eu directive 2003/10/ec (normally called the physical agents directive)". European union. 15 february 2013. Retrieved 26 august 2013

Permissions

All chapters in this book are published with permission under the Creative Commons Attribution Share Alike License or equivalent. Every chapter published in this book has been scrutinized by our experts. Their significance has been extensively debated. The topics covered herein carry significant information for a comprehensive understanding. They may even be implemented as practical applications or may be referred to as a beginning point for further studies.

We would like to thank the editorial team for lending their expertise to make the book truly unique. They have played a crucial role in the development of this book. Without their invaluable contributions this book wouldn't have been possible. They have made vital efforts to compile up to date information on the varied aspects of this subject to make this book a valuable addition to the collection of many professionals and students.

This book was conceptualized with the vision of imparting up-to-date and integrated information in this field. To ensure the same, a matchless editorial board was set up. Every individual on the board went through rigorous rounds of assessment to prove their worth. After which they invested a large part of their time researching and compiling the most relevant data for our readers.

The editorial board has been involved in producing this book since its inception. They have spent rigorous hours researching and exploring the diverse topics which have resulted in the successful publishing of this book. They have passed on their knowledge of decades through this book. To expedite this challenging task, the publisher supported the team at every step. A small team of assistant editors was also appointed to further simplify the editing procedure and attain best results for the readers.

Apart from the editorial board, the designing team has also invested a significant amount of their time in understanding the subject and creating the most relevant covers. They scrutinized every image to scout for the most suitable representation of the subject and create an appropriate cover for the book.

The publishing team has been an ardent support to the editorial, designing and production team. Their endless efforts to recruit the best for this project, has resulted in the accomplishment of this book. They are a veteran in the field of academics and their pool of knowledge is as vast as their experience in printing. Their expertise and guidance has proved useful at every step. Their uncompromising quality standards have made this book an exceptional effort. Their encouragement from time to time has been an inspiration for everyone.

The publisher and the editorial board hope that this book will prove to be a valuable piece of knowledge for students, practitioners and scholars across the globe.

Index